Gareth. S. Aldren

An Introduction to
Environmental Chemistry

An Introduction to Environmental Chemistry

J.E. Andrews, P. Brimblecombe,
T.D. Jickells and P.S. Liss

School of Environmental Sciences,
University of East Anglia,
Norwich NR4 7TJ, UK

**Blackwell
Science**

© 1996 by
Blackwell Science Ltd
Editorial Offices:
Osney Mead, Oxford OX2 0EL
25 John Street, London WC1N 2BL
23 Ainslie Place, Edinburgh EH3 6AJ
238 Main Street, Cambridge
  Massachusetts 02142, USA
54 University Street, Carlton
  Victoria 3053, Australia

Other Editorial Offices:
Arnette Blackwell SA
  224, Boulevard Saint Germain
  75007 Paris, France

Blackwell Wissenschafts-Verlag GmbH
  Kurfürstendamm 57
  10707 Berlin, Germany

  Zehetnergasse 6
  A-1140 Wien,
  Austria

First published 1996
Reprinted 1996

Set by Setrite Typesetters Ltd Hong Kong
Printed and bound in Great Britain by
Hartnolls Ltd, Bodmin, Cornwall

The Blackwell Science logo is a trade mark of
Blackwell Science Ltd, registered at the
United Kingdom Trade Marks Registry

DISTRIBUTORS

Marston Book Services Ltd
PO Box 269
Abingdon
Oxon OX14 4YN
(Orders: Tel: 01235 465500
        Fax: 01235 465555)

USA
Blackwell Science, Inc.
238 Main Street,
Cambridge, MA 02142
(Orders: Tel: 800 215-1000
        617 876-7000
        Fax: 617 492-5263)

Canada
Copp Clark, Ltd
2775 Matheson Blvd East
Mississauga, Ontario
Canada, L4W 4P7
(Orders: Tel: 800 263-4374
        905 238-6074)

Australia
Blackwell Science Pty Ltd
54 University Street
Carlton, Victoria 3053
(Orders: Tel: 03 9347 0300
        Fax: 03 9349 3016)

A catalogue record for this title is available from
the British Library

ISBN 0-632-03854-3

Library of Congress
Cataloging-in-Publication Data

An introduction to environmental
chemistry / J.E. Andrews ... [et al.].
      p.   cm.
  Includes bibliographical references
    (p.   —   ) and index.
  ISBN 0-632-03854-3
  1. Environmental geochemistry.
  I. Andrews, J.E. (Julian E.)
QE516.4.I57   1996
551.9—dc20   95–8320
                                CIP

Contents

List of boxes viii
Preface xi
Acknowledgements xiii
Symbols and abbreviations xvi

1 Introduction 1

1.1 What is environmental chemistry? 1
1.2 In the beginning 1
1.3 Origin and evolution of the Earth 3
    1.3.1 Formation of the crust and atmosphere 3
    1.3.2 The hydrosphere 4
    1.3.3 The origin of life and evolution of the atmosphere 8
1.4 The structure of this book 10
1.5 Further reading 11

2 The atmosphere 12

2.1 Introduction 12
2.2 Composition of the atmosphere 12
2.3 Steady state or equilibrium? 18
2.4 Natural sources 21
    2.4.1 Geochemical sources 21
    2.4.2 Biological sources 24
2.5 Reactivity of trace substances in the atmosphere 27
2.6 The urban atmosphere 30
    2.6.1 London smog — primary pollution 32
    2.6.2 Los Angeles smog — secondary pollution 34
2.7 Air pollution and health 37
2.8 Effects of air pollution 40
2.9 Removal processes 43
2.10 Further reading 45

3 The terrestrial environment 46

3.1 The terrestrial environment, crust and material cycling 46
3.2 The structure of silicate minerals 47
    3.2.1 Coordination of ions and the radius ratio rule 53

3.2.2 The construction of silicate minerals 56
3.2.3 Structural organisation in silicate minerals 56
3.3 Weathering processes 58
3.4 Mechanisms of chemical weathering 59
    3.4.1 Dissolution 59
    3.4.2 Oxidation 59
    3.4.3 Acid hydrolysis 64
    3.4.4 Weathering of complex silicate minerals 65
3.5 Rate controls on weathering reactions 67
    3.5.1 Temperature and water flow rate 67
    3.5.2 Mineral reaction kinetics and solution saturation 69
    3.5.3 Type of parent material (bedrock) 71
    3.5.4 Soils and biology 71
3.6 The solid products of weathering 73
    3.6.1 Clay minerals 73
    3.6.2 Clay mineral composition 73
    3.6.3 One to one clay mineral structure 75
    3.6.4 Two to one clay mineral structure 77
    3.6.5 Controls on clay mineral formation 79
    3.6.6 Ion exchange in soils and the hydrosphere 84
    3.6.7 Use of clay minerals in cases of environmental contamination 85
3.7 The chemistry of continental waters 88
    3.7.1 Element chemistry 89
    3.7.2 Water chemistry and weathering regimes 91
    3.7.3 Silicon and aluminium 96
    3.7.4 Biological processes 98
    3.7.5 Nutrients and eutrophication 101
    3.7.6 Contamination of groundwater 109
3.8 Further reading 113

4 The oceans 114

4.1 Introduction 114
4.2 Estuarine processes 114
    4.2.1 Aggregation of colloidal material in estuaries 114
    4.2.2 Mixing processes in estuaries 116
    4.2.3 Halmyrolysis and ion exchange in estuaries 117
    4.2.4 Microbiological activity in estuaries 118
4.3 Major ion chemistry of seawater 120
4.4 Chemical cycling of major ions 121
    4.4.1 Sea-to-air fluxes 125
    4.4.2 Evaporites 126
    4.4.3 Cation exchange 127

4.4.4 Carbonate precipitation 127
4.4.5 Opaline silica 136
4.4.6 Sulphides 136
4.4.7 Hydrothermal processes 138
4.4.8 Balancing the seawater major ion budget 143
4.4.9 Anthropogenic effects on major ions in seawater 144
4.5 Minor chemical components in seawater 145
4.5.1 Conservative behaviour 146
4.5.2 Nutrient-like behaviour 147
4.5.3 Scavenged behaviour 152
4.5.4 Ocean circulation and its effects on trace element distribution 155
4.6 Further reading 159

5 Global change 160

5.1 Why study global-scale environmental chemistry? 160
5.2 What sorts of substances are involved? 160
5.3 The carbon cycle 162
5.3.1 The atmospheric record 162
5.3.2 Natural and anthropogenic sources and sinks 164
5.3.3 The global budget of natural and anthropogenic carbon dioxide 173
5.3.4 The effects of elevated carbon dioxide levels on global temperature and other properties 178
5.4 The sulphur cycle 182
5.4.1 The global sulphur cycle and anthropogenic effects 182
5.4.2 The sulphur cycle and atmospheric acidity 184
5.4.3 The sulphur cycle and climate 190
5.5 Chlorofluorocarbons and stratospheric ozone 192
5.5.1 Ozone formation and destruction 194
5.5.2 Ozone destruction by chlorine-containing species 195
5.5.3 Limiting chlorofluorocarbon production and searching for alternatives 196
5.6 Further reading 198

Appendices 199
1 Standard electrode potentials at 25°C 199
2 Examples of pH buffering systems 200
Index 203

List of boxes

1.1 Elements, atoms and isotopes 2
1.2 Using chemical equations 9

2.1 The mole 14
2.2 Bonding 14
2.3 Partial pressure 17
2.4 Chemical equilibrium 20
2.5 Acids and bases 23
2.6 Radioactive emanation 25
2.7 Organic molecular structure 26
2.8 Gas solubility 28
2.9 Ozone 30
2.10 The pH scale 35
2.11 Reactions in photochemical smog 38
2.12 Acidification of rain droplets 42
2.13 Removal of sulphur dioxide from an air parcel 44

3.1 Properties of water and hydrogen bonds 48
3.2 Radon gas: a natural environmental hazard 49
3.3 Ionic bonding, ions and ionic solids 51
3.4 Electronegativity 55
3.5 Redox reactions 60
3.6 Reaction kinetics, activation energy and catalysts 62
3.7 Chemical energy 68
3.8 Solubility product, mineral solubility and saturation index 70
3.9 Isomorphous substitution 77
3.10 Van der Waals' forces 80
3.11 Surfactants 89
3.12 Ionic strength 94
3.13 Alkalinity and pH buffering 96
3.14 The Lake Nyos gas disaster: a natural hazard related to lake stratification 102
3.15 Eh-pH diagrams 106

4.1 Salinity 116
4.2 Constancy of major ion chemistry of seawater on geological timescales 122
4.3 Residence times of major ions in seawater 124

4.4   Active concentrations 129
4.5   Ion interactions and ion pairing 130
4.6   Abiological precipitation of calcium carbonate 132
4.7   Hydrothermal circulation at mid-ocean ridges 140
4.8   Oceanic primary productivity 148
4.9   Human effects on regional seas: the Baltic 154

5.1   Simple box model for ocean carbon dioxide uptake 170
5.2   The delta notation for expressing stable isotope ratio values 189

Preface

During the 1980s and 1990s environmental issues have attracted a great deal of scientific, political and media attention. Global and regional-scale issues have received much attention, for example, carbon dioxide (CO_2) emissions linked with global warming, and the depletion of stratospheric ozone by chlorofluorocarbons (CFCs). Local issues, however, have been treated no less seriously, because their effects are more obvious and immediate. The contamination of water supplies by landfill leachate and the build-up of radon gas in domestic dwellings are no longer the property of a few idiosyncratic specialists but the concern of a wide spectrum of the population. It is noteworthy that many of these issues involve understanding chemical reactions and this makes environmental chemistry a particularly important and topical discipline.

We decided the time was right for a new elementary text on environmental chemistry, mainly for students and other readers with little or no previous chemical background. Our aim has been to introduce some of the fundamental chemical principles which are used in studies of environmental chemistry and to illustrate how these apply in various cases, ranging from the global to the local scale. We see no clear boundary between the environmental chemistry of human issues (CO_2 emissions, CFCs, etc.) and the environmental geochemistry of the Earth. A strong theme of this book is the importance of understanding how natural geochemical processes operate and have operated over a variety of timescales. Such an understanding provides baseline information against which the effects of human perturbations of chemical processes can be quantified. We have not attempted to be exhaustive in our coverage but have chosen themes which highlight underlying chemical principles.

We have some experience of teaching environmental chemistry to both chemists and non-chemists through our first-year course in Environmental Chemistry, part of our undergraduate degree in Environmental Sciences at the University of East Anglia. For 14 years we used the text by R.W. Raiswell, P. Brimblecombe, D.L. Dent and P.S. Liss, *Environmental Chemistry*, an earlier University of East Anglia collaborative effort published by Edward Arnold in 1980. The book has served well but is now dated, in part because of the many recent exciting discoveries in environmental chemistry and also partly because the emphasis of the subject has swung toward human concerns and timescales. We have, however, styled parts of the new book on its 'older cousin', particularly where the previous book worked well for our students.

In places the coverage of the present book goes beyond our first-year course and leads on toward honours-year courses. We hope that the material covered will be suitable for other introductory university and college courses in environmental science, earth sciences and geography. It may also be suitable for some courses in life and chemical sciences.

Julian Andrews, Peter Brimblecombe, Tim Jickells and Peter Liss
University of East Anglia, Norwich, UK

Acknowledgements

We would like to thank the following friends and colleagues who have helped us with various aspects of the preparation of this book: Tim Atkinson, Tony Greenaway, Robin Haynes, Kevin Hiscock, Alan Kendall, Gill Malin, Rachel Mills and Willard Pinnock. Special thanks are due to Nicola McArdle for permission to use some of her sulphur isotope data.

We have used or modified tables and figures from various sources, which are quoted in the captions. We thank the various authors and publishers for permission to use this material, which has come from the following sources.

Books

Berner, K.B. & Berner, R.A. (1987) *The Global Water Cycle*. Prentice Hall, Englewood Cliffs.

Berner, R.A. (1980) *Early Diagenesis*. Princeton University Press, Princeton.

Birkeland, P.W. (1974) *Pedology, Weathering, and Geomorphological Research*. Oxford University Press, New York.

Brimblecombe, P. (1986) *Air Composition and Chemistry*. Cambridge University Press, Cambridge.

Broecker, W.S. & Peng, T.-H. (1982) *Tracers in the Sea*. Eldigio Press, New York.

Burton, J.D. & Liss, P.S. (1976) *Estuarine Chemistry*. Academic Press, London.

Garrels, R.M., Mackenzie, F.T. & Hunt, C. (1975) *Chemical Cycles and the Global Environment*. Kaufmann, Los Altos.

Gill, R. (1989) *Chemical Fundamentals of Geology*. Unwin Hyman, London.

IPCC (1990) *Climate Change: The IPCC Scientific Assessment*, ed. by Houghton, J.T., Jenkins, G.J. & Ephramus, J.J. Cambridge University Press, Cambridge.

IPCC (1995) *Radiative Forcing and Climate Change*. Report of the Scientific Assessment Group (WGI) of the Intergovernmental Panel on Climate Change (IPCC) (in press).

Krauskopf, K.B. (1979) *Introduction to Geochemistry*, 2nd edn. McGraw-Hill, Tokyo.

McKie, D. & McKie, C. (1974) *Crystalline Solids*. Nelson, London.

Marland, G. & Boden, T.A. (1989) Carbon Dioxide Releases from Fossil Fuel Burning, testimony before the Senate Committee on Energy and Natural Resourses. 26 July 1989, pp. 62–84, S. Hrg. 101–235, *DOE's National Energy Plan and Global Warming*, US Senate, US Government Printing Office, Washington, D.C.

Moss, B. (1988) *Ecology of Freshwaters*. Blackwell Scientific Publications, Oxford.

Raiswell, R.W., Brimblecombe, P., Dent, D.L. & Liss, P.S. (1980) *Environmental Chemistry*. Edward Arnold, London.

Schaug, *et al.* (1987) *Co-operative Programme for Monitoring and Evaluation of the Long Range Transport of Air Pollutants in Europe (EMEP)*. Summary report of the Norwegian Institute for Air Research, Oslo.

Scoffin, T.P. (1987) *An Introduction to Carbonate Sediments and Rocks*. Blackie, Glasgow.

Spedding, D.J. (1974) *Air Pollution*. Oxford University Press, Oxford.

Strakhov, N.M. (1967) *Principles of Lithogenesis*, vol. 1. Oliver & Boyd, London.

Svedrup, H., Johnson, M.W. & Fleming, R.H. (1941) *The Oceans.* Prentice Hall, Englewood Cliffs.

Taylor, R.S. & McLennan, S.M. (1985) *The Continental Crust: Its Composition and Evolution.* Blackwell Scientific Publications, Oxford.

Wood, L. (1982) *The Restoration of the Tidal Thames.* Adam Higher, Bristol.

Articles

Ayers, G.P., Ivey, J.P. & Gillet, R.W. (1991) *Nature* **349**, 404–406, Macmillan, London.

Boyle, E. A., Collier, R., Dengler, A.T., Edmond, J.M., Ng, A.C. & Stallard, R.F. (1974) *Geochimica Cosmochimica Acta* **38**, 1719–1728, Pergamon, Oxford.

Brimblecombe, P., Hammer, C., Rodhe, H., Ryaboshapko & Boutron, C.F. (1989) in *Evolution of the Global Biogeochemical Sulphur Cycle,* ed. by Brimblecombe, P. & Lein, A. Yu, pp. 77–121, Wiley, Chichester.

Bruland, K.W. (1980) *Earth and Planetary Science Letters* **47**, 189–192, Elsevier, Amsterdam.

Crane, A. & Liss, P.S. (1985) *New Scientist* **108** (1483), 50–54, IPC Magazines, London.

Crawford, N.C. (1984) in *Sinkholes: Their Geology, Engineering and Environmental Impact,* ed. by Beck, B.F., pp. 297–304, Balkema, Rotterdam.

Davies, T.A. & Gorsline, D.S. (1976) in *Chemical Oceanography* vol. 5, ed. by Riley, J.P. & Chester, R., pp. 1–80, Academic Press, London.

Drever, J.I., Li, Y.-H. & Maynard, J.B. (1988) in *Chemical Cycles and the Evolution of the Earth,* ed. by Gregor, C.B., Gregor, C.B., Garrels, R.M., Mackenzie, F.T. & Maynard, J.B. pp. 17–53, Wiley, New York.

Duce *et al.* (1991) The atmospheric input of trace species to the world ocean. *Global Biogeochemical Cycles* **5**, 193–259. American Geophysical Union, Washington, D.C.

Edwards, A. (1973) *Journal of Hydrology* **18**, 219–242, Elsevier, Amsterdam.

Fell, N. & Liss, P.S. (1993) *New Scientist* **139** (1887), 34–38, IPC Magazines, London.

Fichez, R., Jickells, T.D. & Edmonds, H.M. (1992) *Estuarine Coastal Shelf Science* **35**, 577–592, Academic Press, London.

Fonselius, S. (1981) *Marine Pollution Bulletin,* **12**, 187–194, Pergamon, Oxford.

Gibbs, R.J. (1970) *Science* **170**, 1088–1090, American Association for the Advancement of Science, Washington DC.

Gieskes, J.M. & Lawrence, J.R. (1981) *Geochimica Cosmochimica Acta* **45**, 1687–1703, Pergamon, Oxford.

Greenwood, R.J. (1982) *Plant and Soil* **67**, 45–59, Nijhoff/Junk, The Hague.

Houghton, R.A. (1995) in *The Carbon Cycle,* ed. by Schimel, D.S. & Wigley, T.M.L., Cambridge University Press, Cambridge.

Kimmel, G.E. & Braids, O.C. (1980) US Geological Survey Professional Paper 1085, 38 pp, US Government Printing Office, Washington, D.C.

Likens, G., Wright, R.F., Galloway, J.N. & Butler, T.J. (1979) *Scientific American* **241**, 39–47, Scientific American Inc., New York.

Livingstone, D.A. (1963) Chemical composition of rivers and lakes. *US Geological Survey Professional Paper,* US Government Printing Office, Washington, D.C.

Macintyre, I.G. & Reid, R.P. (1992) *Journal of Sedimentary Petrology* **62**, 1095–1097, Society for Sedimentary Geology, Tulsa.

Manabe, S. & Wetherald, R.T. (1980) *Journal of the Atmospheric Sciences* **37**, 99–118, American Meteorological Society, Boston.

Martin, J.-M. & Whitfield, M. (1983) in *Trace Metals in Sea Water,* ed. by Wong C.S., Boyle, E., Bruland, K.W., Burton, J.D. & Goldberg, E.D., Plenum Press, New York.

Martin, R.T., Bailey, S.W., Eberl, D.D. *et al.* (1991) *Clays and Clay Minerals* **39**, 333–335, Clay Minerals Society, Bloomington.

Maynard, J.B., Ritger, S.D. & Sutton, S.J. (1991) *Geology* **19**, 265–268, Geological Society of America, Boulder.

Meybeck, M. (1979) Concentrations des eaux fluviales en élémentes majeurs et appots en solution aux oceans. *Rev Geol Dyn Geogr Phys* **21**(3), 215–246.

Michot, L.J. & Pinnavaia, T.J. (1991) *Clays and Clay Minerals* **39**, 634–641, Clay Minerals Society, Bloomington.

Nehring, D. (1981) *Marine Pollution Bulletin* **12**, 194–198, Pergamon, Oxford.

Nesbitt, H.W. & Young, G.M. (1982) *Nature* **299**, 715–717, Macmillan, London.

Nesbitt, H.W. & Young, G.M. (1984) *Geochimica Cosmochimica Acta* **48**, 1523–1534, Pergamon, Oxford.

Orians, K.J. & Bruland, K.W. (1986) *Earth Planetary Science Letters* **78**, 397–410, Elsevier, Amsterdam.

Rotty, R.M. (1980) *Science of the Total Environment* **15**, 73–86, Elsevier, Amsterdam.

Shen, G.T. & Boyle, E.A. (1987) *Earth Planetary Science Letters* **82**, 289–304, Elsevier, Amsterdam.

Sherman, G.D. (1952) in *Problems in Clay and Laterite Genesis*, p. 154, American Institute of Mining, Metallurgical, Petroleum Engineers, New York.

Sohrin, Y., Isshiki, K. & Kuwamoto, T. (1987) *Marine Chemistry* **22**, 95–103, Elsevier, Amsterdam.

Spilhaus, A.F. (1942) in *Geographical Review* **32**, 431–435, American Geographical Society, New York.

Stallard, R.F. & Edmond J.M. (1983) *Journal of Geophysical Research* **88**, 9671–9688, American Geophysical Union, Washington DC.

Stommel, H. (1958) *Deep Sea Research* **5**, 80–82, Pergamon, London.

Symbols and abbreviations

Multiples and submultiples

Symbol	Name	Equivalent
T	tera	10^{12}
G	giga	10^9
M	mega	10^6
k	kilo	10^3
d	deci	10^{-1}
c	centi	10^{-2}
m	milli	10^{-3}
μ	micro	10^{-6}
n	nano	10^{-9}
p	pico	10^{-12}

Chemical symbols

Symbol	Description	Units
a	activity	mol l^{-1}
c	concentration	mol l^{-1}
eq	equivalents	eq l^{-1}
I	ionic strength	mol l^{-1}
IAP	ion activity product	mol^n l^{-n}
K	equilibrium constant	mol^n l^{-n}
K'	first dissociation constant	mol^n l^{-n}
K_a	equilibrium constant for acid	mol^n l^{-n}
K_b	equilibrium constant for base	mol^n l^{-n}
K_H	Henry's law constant	mol l^{-1} atm^{-1}
K_{sp}	solubility product	mol^n l^{-n}
K_w	equilibrium constant for water	mol^2 l^{-2}
mol	mole (amount of substance — see Box 2.1)	
p	partial pressure	atm

General symbols and abbreviations

Symbol	Description
A	total amount of gas in atmosphere
(aq)	aqueous species
atm	atmosphere (pressure)

$B(\alpha)P$	benzo(α)pyrene
°C	degrees Celsius (temperature)
CCD	calcite compensation depth
CCN	cloud condensation nuclei
CDT	Canyon Diablo troilite
CEC	cation exchange capacity
CFC	chlorofluorocarbon
CIA	chemical index of alteration
D	deuterium
DDT	2,2-di(p-chlorophenyl)-1,1,1-trichloroethane
DIP	dissolved inorganic phosphorus
DMS	dimethyl sulphide
DMSP	beta-dimethylsulphoniopropionate
DNA	deoxyribonucleic acid
$E°$	standard electrode potential (V)
e^-	electron
Eh	redox potential (V)
F	flux
FACE	free-air CO_2 enrichment
G	Gibbs free energy (kJ mol^{-1})
g	gram (weight)
(g)	gas
GtC	gigatonnes expressed as carbon
H	scale height
H	enthalpy (J mol^{-1})
HCFCs	hydrochlorofluorocarbons
$h\nu$	photon of light
IGBP	International Geosphere–Biosphere Programme
J	joule (energy, quantity of heat)
K	kelvin (temperature)
l	litre (volume)
(l)	liquid
ln	natural logarithm
$\log_{10}$	base 10 logarithm
m	metre (length)
M	a third body
MSA	methanesulphonic acid
N	neutron number
n	an integer
Pa	pascal (pressure)
PAN	peroxyacetylnitrate
PAH	polycyclic aromatic hydrocarbon
ppb	parts per 10^9
ppm	parts per million

PCBs	polychlorinated biphenyls
r	ionic radius
S	entropy ($J\ mol^{-1}\ K^{-1}$)
s	second (time)
(s)	solid
T	absolute temperature (kelvin)
TBT	tributyl tin
UV	ultraviolet radiation
V	volt (electrical potential)
V	volume
W	watt (power — Js^{-1})
wt%	weight percent
Z	atomic number
z	charge
$\lvert z \rvert$	charge ignoring sign

Greek symbols

α	alpha particle (radiation)
γ	activity coefficient
γ	gamma particle (radiation)
δ	stable isotope notation (Box 5.2)
$\delta-$	partial negative charge
$\delta+$	partial positive charge
Δ	change in
Σ	sum of
τ	residence time
Ω	degree of saturation

Constants

F	Faraday constant ($6.02 \times 10^{23}\ e^-$)
R	gas constant ($8.314\ J\ mol^{-1}\ K^{-1}$)

1 Introduction

1.1 What is environmental chemistry?

It is probably true to say that the term *environmental chemistry* has no precise definition. It means different things to different people. We are not about to offer a new definition. It is clear that environmental chemists are playing their part in the big environmental issues — stratospheric ozone (O_3) depletion, global warming and the like. Similarly, the role of environmental chemistry in regional-scale and local problems — for example, the effects of acid rain or contamination of water resources — is well established. This brief discussion illustrates the clear link in our minds between environmental chemistry and human beings. For many people, 'environmental chemistry' is implicitly linked to 'pollution'. We hope this book demonstrates that such a view is limited and shows that 'environmental chemistry' has a much wider scope.

Terms like *contamination* and *pollution* have little meaning without a frame of reference for comparison. How can we hope to understand the behaviour and impacts of chemical contaminants without understanding how natural chemical systems work? For many years a relatively small group of scientists has been steadily unravelling how the chemical systems of the Earth work, both today and in the geological past. The discussions in this book draw on a small fraction of this material. Our aim is to demonstrate the various scales, rates and types of natural chemical processes that occur on Earth. We also attempt to show the actual or possible effects that humans may have on natural chemical systems. The importance of human influences is usually most clear when direct comparison with the unperturbed, natural systems is possible.

This book deals mainly with the Earth as it is today, or as it has been over the last few million years, with the chemistry of water on the planet's surface a recurrent theme. This theme emphasises the link between natural chemical systems and organisms, not least human beings, since water is the key compound in sustaining life itself. We will start by explaining how the main components of the near-surface Earth — the crust, oceans and atmosphere — originated and how their broad chemical composition evolved. Since all chemical compounds are built from atoms of individual elements (Box 1.1), we begin with the origin of these fundamental chemical components.

1.2 In the beginning

It is believed that the universe began at a single instant in an enormous explosion, often called the *big bang*. Astronomers still find evidence of this explosion in the

Box 1.1

Elements, atoms and isotopes

Elements are made from atoms — the smallest particle of an element that can take part in chemical reactions. Atoms have three main components: protons, neutrons and electrons. Protons are positively charged, with a mass similar to that of the hydrogen atom. Neutrons are uncharged and of equal mass to protons. Electrons are about 1/1836 the mass of protons, with a negative charge of equal value to the (positive) charge of protons.

Atoms are electrically neutral because they have an equal number (Z) of protons and electrons. Z is known as the atomic number and it characterises the chemical properties of the element.

The atomic weight of an atom is defined by its mass number and most of the mass is present in the nucleus.

Mass number = number of protons (Z) + number of neutrons (N)   eq. 1

Equation 1 shows that the mass of an element can be changed by altering the number of neutrons. This does not affect the chemical properties of the element (which are determined by Z). Atoms of an element which differ in mass (i.e. N) are called isotopes. For example, all carbon atoms have a Z number of 6, but mass numbers of 12, 13 and 14, written:

^{12}C, ^{13}C, ^{14}C (isotopes of carbon)

In general, when the number of protons and neutrons in the nucleus are almost the same (i.e. differ by one or two), the isotopes are stable. As Z and N numbers become more dissimilar, isotopes tend to be unstable and break down by radioactive decay (usually liberating heat) to a more stable isotope. Unstable isotopes are called radioactive isotopes.

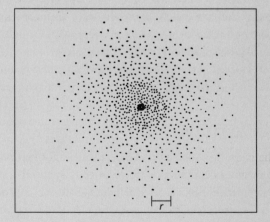

Fig. 1 Representation of the hydrogen atom. The dots represent the position of the electron with respect to the nucleus. The electron moves in a wave motion. It has no fixed position relative to the nucleus, but the *probability* of finding the electron at a given radius (the Bohr radius, r) can be calculated; $r = 5.3 \times 10^{-5}$ μm for hydrogen.

movement of galaxies and the microwave background radiation once associated with the primeval fireball. In the first fractions of a second after the big bang, the amount of matter and radiation, at a ratio of about 1 in 10^8, was fixed. Minutes later the relative abundances of hydrogen (H), deuterium (D) and helium (He) were determined. Heavier elements had to await the formation and processing of these gases within stars. Elements as heavy as iron (Fe) can be made in the cores of stars, while stars which end their lives as explosive supernovae can produce much heavier elements.

Hydrogen and helium are the most abundant elements in the universe, relics of the earliest moments in element production. However, it is the stellar production process that led to the characteristic cosmic abundance of the elements (Fig. 1.1). Lithium (Li), beryllium (Be) and boron (B) are not very stable in stellar interiors, hence the low abundance of these light elements in the universe. Carbon (C), nitrogen (N) and oxygen (O) are formed in an efficient cyclic process in stars that leads to their relatively high abundance. Silicon (Si) is rather resistant to photodissociation (destruction by light) in stars, so it is also abundant and dominates the rocky world we see about us.

1.3 Origin and evolution of the Earth

The planets of our solar system probably formed from a disc-shaped cloud of hot gases, the remnants of a stellar supernova. Condensing vapours formed solids which coalesced into small bodies (planetesimals) and accretion of these built the dense inner planets (Mercury to Mars). The larger outer planets, being more distant from the sun, are composed of lower-density gases, which condensed at much cooler temperatures.

As the early Earth accreted to something like its present mass, it heated up, mainly due to the radioactive decay of unstable isotopes (Box 1.1) and partly by trapping kinetic energy from planetesimal impacts. This heating melted iron and nickel (Ni) and their high densities allowed them to sink to the centre of the planet, forming the core. Subsequent cooling allowed solidification of the remaining material into the mantle of $MgFeSiO_3$ composition (Fig. 1.2).

1.3.1 Formation of the crust and atmosphere

The crust, hydrosphere and atmosphere formed mainly by release of materials from within the upper mantle of the early Earth. Today, ocean crust forms at mid-ocean ridges, accompanied by the release of gases and small amounts of water. Similar processes probably accounted for crustal production on the early Earth, forming a shell of rock less than 0.0001% of the volume of the whole planet (Fig. 1.2). The composition of this shell, which makes up the continents and ocean crust, has evolved over time, essentially distilling elements from the mantle by partial melting at about 100 km depth. The average chemical composition of the present

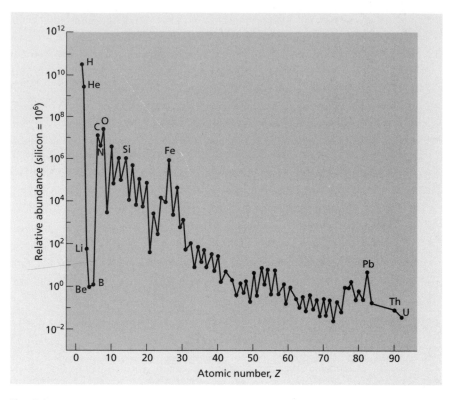

Fig. 1.1 The cosmic abundance of elements. The relative abundance of elements (vertical axis) is defined as the number of atoms of each element per 10^6 atoms of silicon and is plotted on a logarithmic scale.

crust (Fig. 1.3) shows that oxygen is the most abundant element, combined in various ways with silicon, aluminium (Al) and other elements to form silicate minerals.

Various lines of evidence suggest that volatile elements escaped (degassed) from the mantle by volcanic eruptions associated with crust building. Some of these gases were retained to form the atmosphere once surface temperatures were cool enough and gravitational attraction was strong enough. The primitive atmosphere was probably composed of carbon dioxide (CO_2) and nitrogen gas (N_2) with some hydrogen and water vapour. Evolution toward the modern oxidising atmosphere did not occur until life began to develop.

1.3.2   The hydrosphere

Water, in its three phases, liquid water, ice and water vapour, is highly abundant at the Earth's surface, having a volume of 1.4 billion km^3. Nearly all of this water (>97%) is stored in the oceans, while most of the rest forms the polar ice-caps and glaciers (Table 1.1). Continental freshwaters represent less than 1% of the

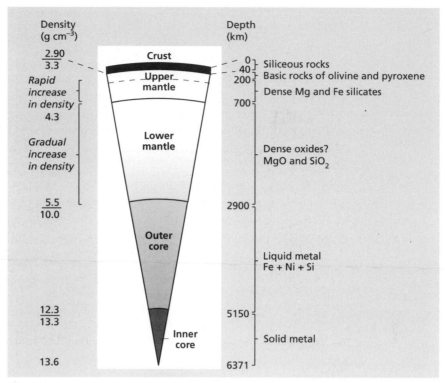

Fig. 1.2 Schematic cross-section of the Earth. Silica is concentrated in the crust relative to the mantle. After Raiswell *et al.* (1980).

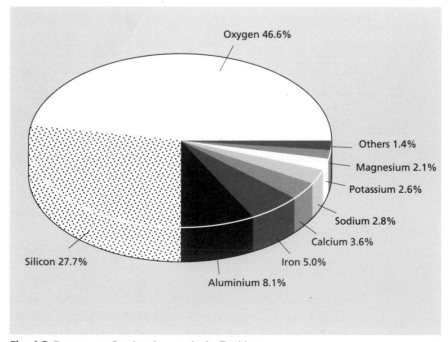

Fig. 1.3 Percentage of major elements in the Earth's crust.

total volume, and most of this is groundwater. The atmosphere contains comparatively little water (as vapour) (Table 1.1). Collectively, these reservoirs of water are called the *hydrosphere*.

The source of water for the formation of the hydrosphere is problematic. Some meteorites contain up to 20% water in bonded hydroxyl (OH) groups, while bombardment of the proto-Earth by comets rich in water vapour is another possible source. Whatever the origin, once the Earth's surface cooled to 100°C, water vapour, degassing from the mantle, was able to condense. We know from the existence of sedimentary rocks laid down in water that the oceans had formed by 3.8×10^9 years ago.

Very little water vapour escapes from the atmosphere to space because, at about 15 km height, the low temperature causes the vapour to condense and fall to lower levels. It is also thought that very little water degasses from the mantle today. These observations suggest that, after the main phase of degassing, the total volume of water at the Earth's surface changed little over geological time.

Cycling between reservoirs in the hydrosphere is known as the *hydrological cycle* (shown schematically in Fig. 1.4). Although the volume of water vapour contained in the atmosphere is small, water is constantly moving through this reservoir. Water evaporates from the oceans and land surface and is transported within air masses. Despite a short residence time (see Section 2.3) in the atmosphere, typically 10 days, the average transport distance is about 1000 km. The water vapour is then returned to either the oceans or the continents as snow or rain. Most rainfall on to the continents seeps into sediments and porous or fractured rock to

Table 1.1 Inventory of water at the Earth's surface. After Berner E.K. & Berner R.A., *The Global Water Cycle: Geochemistry and Environment*, © 1987, p. 13. Reprinted by permission of Prentice Hall, Englewood Cliffs, New Jersey

Reservoir	Volume (10^6 km^3)	Percentage of total
Oceans	1370	97.25
Ice-caps and glaciers	29	2.05
Deep groundwater (750–4000 m)	5.3	0.38
Shallow groundwater (< 750 m)	4.2	0.30
Lakes	0.125	0.01
Soil moisture	0.065	0.005
Atmosphere*	0.013	0.001
Rivers	0.001 7	0.000 1
Biosphere	0.000 6	0.000 04
Total	1408.7	100

* As liquid equivalent of water vapour.

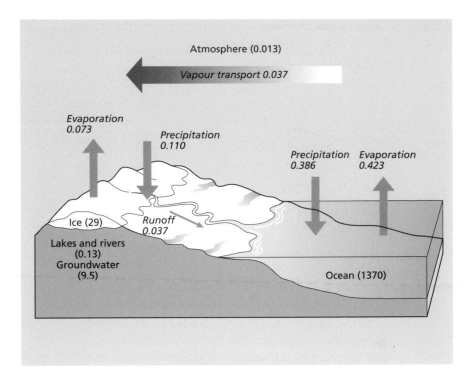

Fig. 1.4 Schematic diagram of the hydrological cycle. Numbers in parentheses are reservoir inventories (10^6 km^3). Fluxes are in 10^6 km^3 year^{-1}. After Berner E.K. & Berner R.A., *The Global Water Cycle: Geochemistry and Environment*, © 1987, p. 14. Reprinted by permission of Prentice Hall, Englewood Cliffs, New Jersey.

form groundwater; the rest flows on the surface as rivers, or re-evaporates to the atmosphere. Since the total mass of water in the hydrosphere is relatively constant over time, evaporation and precipitation must balance for the Earth as a whole, despite locally large differences between wet and arid regions.

The rapid transport of water vapour in the atmosphere is driven by incoming solar radiation. Almost all of the radiation which reaches the crust is used to evaporate liquid water to form atmospheric water vapour. The energy used in this transformation, which is then held in the vapour, is called *latent heat*. Most of the remaining radiation is absorbed into the crust with decreasing efficiency with increasing latitude, mainly because of the Earth's spherical shape. Solar rays hit the Earth's surface at 90° at the equator, but at decreasing angles with increasing latitude, approaching 0° at the poles. Thus, a similar amount of radiation is spread over a larger area at higher latitudes compared with the equator (Fig. 1.5). The variation of incoming radiation with latitude is not balanced by an opposite effect for radiation leaving the Earth, so the result is an overall radiation imbalance. The poles, however, do not get progressively colder and the equator warmer, because heat moves poleward in warm ocean currents and there is poleward movement of warm air and latent heat (water vapour).

1.3.3   The origin of life and evolution of the atmosphere

We do not know which chance events brought about the synthesis of organic molecules or the assembly of metabolising, self-replicating structures which we call organisms, but we can guess at some of the requirements and constraints. In the 1950s there was considerable optimism that the discovery of deoxyribonucleic acid (DNA) and the laboratory synthesis of likely primitive biomolecules from experimental atmospheres rich in methane (CH_4) and ammonia (NH_3) indicated a clear picture for the origin of life. However, it now seems more likely that the synthesis of biologically important molecules occurred in restricted, specialised environments, such as the surfaces of clay minerals, or in submarine volcanic vents.

Best guesses suggest that life began in the oceans some 4.2–3.8 billion years ago, but there is no fossil record. The oldest known fossils are bacteria, some 3.5 billion years old. In rocks of this age there is fossil evidence of quite advanced metabolisms which utilised solar energy to synthesise organic material. The earliest of these reactions were probably based on sulphur (S), supplied from volcanic vents,

$$CO_{2(g)} + 2H_2S_{(g)} \rightarrow \underset{\substack{\text{(organic}\\\text{matter)}}}{CH_2O_{(s)}} + 2S_{(s)} + H_2O_{(l)} \qquad \text{eq. 1.1}$$

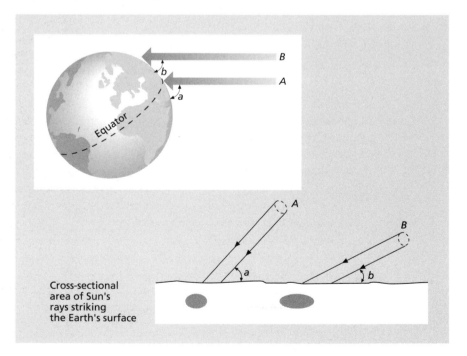

Fig. 1.5 Variation in relative amounts of solar radiation (energy per unit area) with latitude. Equal amounts of energy A and B are spread over a larger area at higher latitude, resulting in reduced intensity of radiation.

but were ultimately achieved by the photochemical splitting of water, or photosynthesis.

$$H_2O_{(l)} + CO_{2(g)} \rightarrow CH_2O_{(s)} + O_{2(g)} \qquad \text{eq. 1.2}$$

(If you are unfamiliar with chemical reactions and notation see Box 1.2.)

The production of oxygen during photosynthesis had a profound effect. Initially, the oxygen gas (O_2) was rapidly consumed, oxidising reduced compounds and minerals. However, once the rate of supply exceeded consumption, O_2 began to build up in the atmosphere. The primitive biosphere, mortally threatened by its own poisonous by-product (O_2), was forced to adapt to this change. It did so by evolving new biogeochemical metabolisms which support the diversity of life on

---

Box 1.2

Using chemical equations

The chemical principles discussed in this book are often illustrated using equations. It is useful to know a few of the ground rules we have adopted to construct these. Let us begin by looking at an equation which depicts the process of rusting metallic iron.

$$4Fe_{(s)} + 3O_{2(g)} \rightarrow 2Fe_2O_{3(s)} \qquad \text{eq. 1}$$

Firstly, the arrow shows that the reaction is favoured in one direction (we will demonstrate this later when discussing energy needed to drive reactions). Next we can see that the reaction *balances*, i.e. we have four atoms of iron and six atoms of oxygen on both sides of the equation. When chemical reactions take place, we neither gain nor lose atoms. Finally, the subscripted characters in brackets represent the *status* of the chemical species. In this book l = liquid, g = gas, s = solid and aq = an aqueous species, i.e. a component dissolved in water.

It is important to realise that these reactions are usually simplifications of the actual chemical transformations that occur in nature. In eq. 1 we are representing rusted or oxidised iron as Fe_2O_3, the mineral haematite. In nature, rusted metal is a complex mixture of iron hydroxides and water molecules. So reaction 1 summarises a series of complicated reaction stages. It illustrates a product we might reasonably expect to form without necessarily depicting the stages of reaction or the complexity encountered in nature.

Many of the equations in this book are written with the reversible reaction sign (two-way half-arrows — see eq. 2). This shows that the reaction can proceed in either direction and this is fundamental to equilibrium-based chemistry (see Box 2.4). Reactions depicting dissolution of substances in water may or may not show the water molecule involved, but dissolution is implied by the (aq) status symbol. Equation 2, read from left to right, shows dissolution of rock salt (halite).

$$NaCl_{(s)} \rightleftharpoons Na^+_{(aq)} + Cl^-_{(aq)} \qquad \text{eq. 2}$$

The reverse reaction (right to left) shows crystallisation of salt from solution.

the modern Earth. Gradually an atmosphere of modern composition evolved (see Table 2.1). In addition, oxygen in the stratosphere (see Chapter 2) underwent photochemical reactions, leading to the formation of ozone (O_3), protecting the Earth from ultraviolet radiation. This shield allowed higher organisms to colonise the continental land surfaces.

In recent decades a few scientists have argued that the Earth acts like a single living entity rather than a randomly driven geochemical system. There has been much philosophical debate about this issue, often called the Gaia hypothesis, and more recently, Gaia theory. This view, suggested by James Lovelock, argues that biology controls the habitability of the planet, making the atmosphere, oceans and terrestrial environment comfortable to sustain and develop life. There is little consensus about these Gaian notions, but the ideas of Lovelock and others have stimulated active debate about the role of organisms in mediating geochemical cycles. Many scientists use the term 'biogeochemical cycles', which acknowledges the role of organisms in influencing geochemical systems.

1.4 The structure of this book

In the following chapters we describe how components of the Earth's chemical systems operate. The emphasis in each chapter is different, reflecting the wide range of chemical compositions and rates of reactions that occur in near-surface Earth environments. The modern atmosphere (Chapter 2), where rates of reaction are rapid, is strongly influenced by human activities. In terrestrial environments (Chapter 3), a huge range of solid and fluid processes interact. The emphasis here is on weathering processes and their influence on the chemical composition of sediments, soils and continental surface waters. The weathering theme links with the oceans (Chapter 4), but it soon becomes clear that the chemical composition of this vast water reservoir is controlled by a host of other physical, biological and chemical processes. Chapter 5 examines environmental chemistry on a global scale, integrating information from earlier chapters and, in particular, focusing on the influence of humans on global chemical processes. The short-term carbon and sulphur cycles are examples of natural chemical cycles perturbed by human activities. In contrast, the reaction between chlorofluorocarbons (CFCs) and stratospheric O_3 is discussed as an example of an unforeseen impact on the natural environment caused by chemicals synthesised by humans.

In all of these chapters we have chosen subjects which demonstrate the chemical principles involved. For those unfamiliar with basic chemistry we provide information boxes which describe, in simplified terms, some of the laws, assumptions and techniques used by chemists. In addition, we have provided boxes outlining case-studies or specific examples, and often these describe chemical impacts caused by humans.

1.5   **Further reading**

Allegre, C. (1992) *From Stone to Star*. Harvard University Press, Cambridge, Massachusetts.

Broecker, W.S. (1985) *How to Build a Habitable Plane*t. Lamont-Doherty Geological Observatory, Columbia University, Palisades, New York.

Emiliani, C. (1992) *Planet Earth: Cosmology, Geology and the Evolution of Life and Environment*. Cambridge University Press, New York.

Lovelock, J. (1982) *Gaia: A New Look At Life on Earth*. Oxford University Press, Oxford.

Lovelock, J. (1988) *The Ages of Gaia*. Oxford University Press, Oxford.

2 The atmosphere

2.1 Introduction

The atmosphere is in the news! Atmospheric chemistry has become a matter of public concern in the last two decades. While the complexities of modern science do not usually spark off great political and social debate, the changes in the atmosphere have evoked great interest. Heads of state have been forced to meetings in Stockholm, Montreal, London and Rio de Janeiro and given their attention to the fate of our atmosphere. Television, which normally relegates scientific matters to off-peak hours, has shown skilfully created colourful images from remotely sensed measurements of the ozone (O_3) hole and huge emissions from oil fires during the 1991 Gulf war. What has caused this interest in the atmosphere?

The atmosphere is the smallest of the Earth's geological reservoirs (Fig. 2.1). It is this limited size that makes the atmosphere potentially so vulnerable to contamination. Even the addition of a small amount of material can lead to significant changes in the way the atmosphere behaves.

We should note that the mixing time of the atmosphere is very rapid. Debris from a large accident, such as the one at the nuclear reactor at Chernobyl in 1986, can quickly be detected all over the globe. This mixing, while distributing contaminants widely, dilutes them at the same time. In contrast, the spread of contaminants in the ocean is much slower and in the other reservoirs of the Earth takes place only over geological timescales of millions of years.

2.2 Composition of the atmosphere

Bulk composition of the atmosphere is quite similar all over the Earth because of the high degree of mixing within the atmosphere. This mixing is driven in a horizontal sense by the rotation of the Earth. Vertical mixing is largely the product of heating of the surface of the Earth by incoming solar radiation. The oceans have a much slower mixing rate, but even this is sufficient to ensure a relatively constant bulk composition in much the same way as the atmosphere. However, some parts of the atmosphere are not so well mixed and here quite profound changes in bulk composition are found.

The lower atmosphere, which is termed the troposphere, is well mixed by convection. Thunderstorms are the most apparent of the convective driving forces. Temperature declines with height in the troposphere; solar energy heats the surface of the Earth and this in turn heats the directly overlying air, causing the convective mixing. This is because the warmer air that is in contact with the surface of the Earth is lighter and tends to rise. However, at a height of some 15–25 km, the

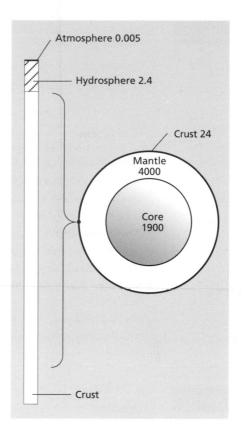

Fig. 2.1 Relative sizes of the major reservoirs of the Earth. Units, 10^{24} g.

atmosphere is heated by the absorption of ultraviolet radiation by oxygen (O_2) and O_3. The rise in temperature with height has the effect of giving the upper part of the atmosphere great stability against vertical mixing. This is because the heavy cold air at the bottom has no tendency to rise. This region of the atmosphere has air in distinct layers or strata and is thus called the stratosphere. The well-known O_3 layer forms at these altitudes. Despite this stability, the stratosphere is comparatively well mixed compared with the atmosphere even higher up. Above about 120 km, turbulent mixing is so weak that individual gas molecules can separate under gravitational settling. Thus the relative concentrations (Box 2.1) of atomic oxygen (O) and nitrogen (N) (Box 2.2) are greatest lower down, while the lighter hydrogen (H) and helium (He) are found to predominate higher up.

Figure 2.2 shows the various layers of the atmosphere. The part where gravitational settling occurs is usually termed the heterosphere, because of the varying composition. The better-mixed part of the atmosphere below is called the homosphere. Turbopause is the term given to the boundary that separates these two parts. The heterosphere is so high (hundreds of kilometres) that pressure is extremely low, as emphasised by the logarithmic scale in the figure.

Box 2.1

The mole

Chemists have adopted a special unit of measurement called the *mole* (abbreviation mol) to describe the amount of substance. A mole is defined as $6.022\ 136\ 7 \times 10^{23}$ molecules. This is chosen to be equivalent to its molecular weight in grams. Thus 1 mol of sodium, which has an atomic weight of 23 (if one were to be very accurate, it would be 22.9898), weighs 23 g and contains $6.022\ 136\ 7 \times 10^{23}$ molecules. This special number of particles is called the Avogadro number, in honour of the Italian physicist Amedeo Avogadro.

The mole is used because it always refers to the same amount of substance, in terms of the number of molecules, regardless of mass of the atoms involved.

Units of concentration are, for this reason, also expressed in terms of the mole. Thus a concentration is given as the amount of substance per unit volume as $mol\ dm^{-3}$ (the unit of molarity) or per unit weight as $mol\ kg^{-1}$ (molality). The latter is now used frequently in chemistry because it has a number of advantages (such as it does not depend on temperature). However, at 25°C, in a dilute solution, $mol\ kg^{-1}$, $mol\ dm^{-3}$ and $mol\ l^{-1}$ are almost equivalent. Although $mol\ l^{-1}$ is not an accepted SI unit, it remains widely used in the environmental sciences, which is the reason we have decided to use it throughout this book.

Box 2.2

Bonding

Many elements do not normally exist as atoms, but are bonded together to form molecules. The major components of air, oxygen and nitrogen are, for example, present in the lower atmosphere as the molecules O_2 and N_2. In contrast, argon is rather unusual because as an inert element (or noble gas) it is found uncombined as single argon atoms. Inert elements are exceptions and most substances in the environment are in the form of molecules.

Molecular bonds are formed from the electrostatic interactions between electrons and the nuclei of atoms. There are many different electronic arrangements that lead to bond formation and the type of bond formed influences the properties of the compound that results. It is the outermost electrons of an atom that are involved in bond formation. The archetypical chemical bond is the covalent bond and we can probably best imagine this as formed from outer electrons shared between two atoms. Take the example of two fluorine atoms which form the fluorine molecule.

$$\overset{..}{:}\underset{..}{F} + \overset{..}{:}\underset{..}{F} \rightleftharpoons \overset{..}{:}\underset{..}{F} : \overset{..}{F} :$$

In this representation of bonding, the electrons are shown by dots. In reality the bonding electrons are smeared out over the entire molecules, but their most probable position is between the nuclei. The bond is shown by the two electrons

**Box 2.2
Cont.**

between the atoms. The bond is created from the two electrons shared between the atoms. In simple terms it can be argued that this arrangement of electrons achieves a structure which is similar to that of argon.

$$: \overset{..}{\underset{..}{Ar}} :$$

Thus bond formation can be envisaged as a result of attaining noble-gas-type structures, which have a particularly stable configurations of electrons. Symbolically this covalent bond is written F-F. We can think of the bonding electrons, which tend to sit between the two nuclei, as shielding the repulsive forces of the protons in the nucleus.

Oxygen and nitrogen are a little different.

$$: \overset{\cdot}{\underset{..}{O}}. \ + \ : \overset{\cdot}{\underset{..}{O}}. \ \rightleftharpoons \ : \overset{..}{O} :: \overset{..}{O} :$$

Here the argon-like structure requires two electrons from each atom and the double bond formed is symbolised O = O. For nitrogen we have:

$$: \overset{\cdot}{\underset{\cdot}{N}}. \ + \ : \overset{\cdot}{\underset{\cdot}{N}}. \ \rightleftharpoons \ : N ::: N :$$

symbolised $N \equiv N$ (a triple bond).

Gases in the atmosphere, water and organic compounds are typically formed with these kinds of covalent bonds.

In a mixture of gases like the atmosphere, Dalton's law of partial pressure (Box 2.3) is obeyed. This means that individual gases in the atmosphere will decline in pressure at the same rate as the total pressure. This can all be conveniently represented by the barometric equation:

$$p_z = p_0 \exp(-z/H) \qquad \qquad \text{eq. 2.1}$$

where p_z is the pressure at altitude z, p_0 the pressure at ground level and H, the scale height (about 8.4 km in the lower troposphere and a measure of the rate at which pressure falls with height). We can see from this equation that the pressure declines so rapidly in the lower atmosphere that it reaches 50% of its ground level value by 5.8 km. This is painfully obvious to people who have found themselves exhausted when trying to climb high mountains. We should note that if eq. 2.1 is integrated over the troposphere it accounts for about 90% of all atmospheric gases. The rest are largely in the stratosphere and the low mass of the upper atmosphere reminds us that it will be sensitive to pollutants. There is so little gas in the stratosphere that relatively small amounts of trace pollutants can have a big impact.

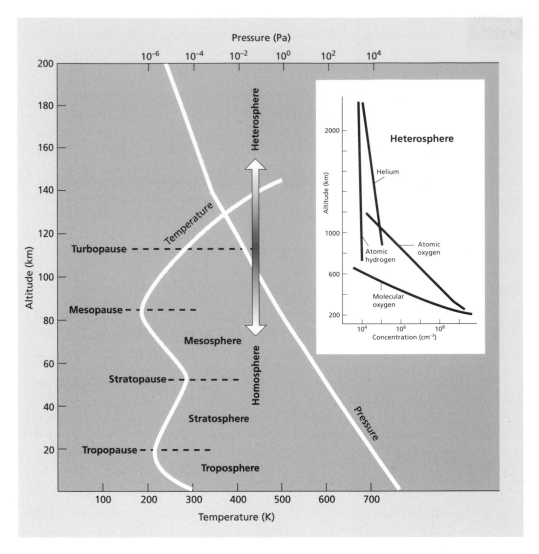

Fig. 2.2 The vertical structure of the atmosphere and associated temperature and pressure variation. Note the logarithmic scale for pressure. The inset shows gas concentration as a function of height in the heterosphere and illustrates the presence of lighter gases (hydrogen and helium) at greater heights.

Furthermore, pollutants will be held in relatively well-defined layers because of the restricted vertical mixing and this will prevent dispersal and dilution.

It is well known that the atmosphere consists mostly of nitrogen (N_2) and O_2, with a small percentage of argon (Ar). The concentrations of the major gases are listed in Table 2.1. Water (H_2O) is also an important gas, but its abundance varies a great deal. In the atmosphere as a whole, the concentration of water is dependent on temperature. Carbon dioxide (CO_2) has a much lower concentration, as do many other relatively inert (i.e. unreactive) trace gases. Apart from water, and to a lesser extent CO_2, most gases remain at fairly constant concentrations in the atmosphere.

Box 2.3

Partial pressure

The total pressure of a mixture of gases is equal to the sum of the pressures of the individual components. The pressure–volume relationship of an ideal gas (i.e. a gas composed of atoms with negligible volume and which undergo perfectly elastic collisions with one another) is defined as:

$$pV = nRT \qquad \text{eq. 1}$$

where p is the partial pressure, V the volume, n the number of moles of gas, R the gas constant and T the absolute temperature. Real gases behave like ideal gases at low pressure and we denote a mixture of gases (1, 2, 3...) through the use of subscripts:

$$p_1 V = n_1 RT$$
$$p_2 V = n_2 RT$$
$$p_3 V = n_3 RT$$

Hence:

$$(p_1 + p_2 + p_3)\, V = (n_1 + n_2 + n_3)RT \qquad \text{eq. 2}$$

or

$$p_T V = (n_1 + n_2 + n_3)RT \qquad \text{eq. 3}$$

where p_T is the total pressure of the mixture. The implication that partial pressure p_i is a function of n_i means that the barometric law (p. 15):

$$p_z = p_0 \exp(-z/H) \qquad \text{eq. 4}$$

can be rewritten:

$$n_z = n_0 \exp(-z/H) \qquad \text{eq. 5}$$

or even:

$$X_z = X_0 \exp(-z/H) \qquad \text{eq. 6}$$

where X is some unit of the amount of material per unit volume (g m^{-3} or molecules cm^{-3}). The barometric law predicts that pressure and concentration of gases in the atmosphere decline at the same rate with height.

The relationship between partial pressure and gas phase concentration explains why concentrations in the atmosphere are frequently expressed in parts per million (ppm) or parts per billion (ppb) (see Table 2.1). This is done on a volume basis so that 1 ppm means 1 cm^3 of a substance is present in 10^6 cm^3 of air. It also requires that there is one molecule of the substance present for every million molecules of air, or one mole of the substance present for every million moles of air. This ppm unit is thus a kind of mole ratio. It can be directly related to pressure through the law of partial pressure, so at one atmosphere (1 atm) pressure a gas present at a concentration of 1 ppm will have a pressure of 10^{-6} atm.

Table 2.1 Bulk composition of unpolluted air. These are the components that provide the background medium in which atmospheric chemistry takes place. From Brimblecombe (1986)

Gas	Concentration
Nitrogen	78.084%
Oxygen	20.946%
Argon	0.934%
Water	0.5–4%
Carbon dioxide	360 ppm
Neon	18.18 ppm
Helium	5.24 ppm
Methane	1.7 ppm
Krypton	1.14 ppm
Hydrogen	0.5 ppm
Xenon	0.087 ppm

Although these non-variant gases can hardly be said to be unimportant, the attention of atmospheric chemists usually focuses on the reactive trace gases. In the same way, much interest in the chemistry of seawater revolves around its trace components and not water itself or sodium chloride (NaCl), its main dissolved salt.

2.3 **Steady state or equilibrium?**

Let us look at an individual trace gas in the atmosphere. We will take methane (CH_4), not an especially reactive gas, as an illustration. It is present in the atmosphere at about 1.7 ppm (see Box 2.3). Methane can react with O_2 in the following way:

$$CH_{4\,(g)} + 2O_{2\,(g)} \rightarrow CO_{2\,(g)} + 2H_2O_{\,(g)} \qquad \text{eq. 2.2}$$

The reaction can be represented as an equilibrium situation (Box 2.4) and described by the conventional equation:

$$K = \frac{cCO_2.cH_2O^2}{cCH_4.cO_2^2} \qquad \text{eq. 2.3}$$

which can be written in terms of pressure (see Box 2.2):

$$K = \frac{pCO_2.pH_2O^2}{pCH_4.pO_2^2} \qquad \text{eq. 2.4}$$

The equilibrium constant (K) is about 10^{140} (Box 2.4). This is an extremely large number, which suggests that the equilibrium position of this reaction lies very much to the right and that CH_4 should tend to be at low concentrations in the atmosphere. How low? We can calculate this by rearranging the equation and solving for CH_4. Oxygen, we can see from Table 2.1, has a concentration of about 21%, i.e. 0.21 atm, while CO_2 and water have values of 0.000 36 and about 0.01 atm respectively. Substituting these into eq. 2.4 and solving the equation gives an equilibrium concentration of 8×10^{-147} atm. This is very different from the value of 1.7×10^{-6} atm actually found present in air.

What has gone wrong? This simple calculation tells us that gases in the atmosphere are not necessarily in equilibrium. This does not mean that the atmosphere is especially unstable, but just that it is not governed by chemical equilibrium. Many trace gases in the atmosphere are in steady state. Steady state describes the delicate balance between the input and output of the gas to the atmosphere. The notion of a balance between the source of a gas to the atmosphere and sinks for that gas is an extremely important one. The situation is often written in terms of the equation:

$$F_{in} = F_{out} = \frac{A}{\tau}$$
eq. 2.5

where F_{in} and F_{out} are the fluxes in and out of the atmosphere, A is the total amount of the gas in the atmosphere and τ is the residence time of the gas.

To be in steady state the input term must equal the output term. Imagine the atmosphere as a leaky bucket into which a tap is pouring water. The bucket would fill for a while until the pressure rose and the leaks were rapid enough to match the inflow rate. At that point we could say that the system was in steady state.

Methane input into the atmosphere occurs at a rate of 500 Tg year^{-1} (i.e. 500×10^9 kg year^{-1}). We have seen that the atmosphere has CH_4 at a concentration of 1.7 ppm. The total atmospheric mass is 5.2×10^{18} kg. If we allow for the slight differences between the molecular mass of CH_4 and that of the atmosphere as a whole (i.e. 16/29), the total mass of CH_4 in the atmosphere can be estimated as 4.8×10^{12} kg. Substituting these values in eq. 2.5 gives a residence time of 9.75 years. This represents the average lifetime of a CH_4 molecule in the atmosphere (at least, it would if the atmosphere was very well mixed).

Residence time is the fundamental quantity that describes systems in steady state. It is a very powerful concept that plays a central role in much of environmental chemistry. Compounds with long residence times can accumulate to relatively high concentrations compared with those with shorter ones. However, even though gases with short residence times are removed quickly, their high reactivity can yield reaction products that cause problems.

The famous atmospheric chemist C.E. Junge made an important observation about residence times and the variability of gases in the atmosphere. If a gas has a long residence time, then it will have ample time to become well mixed in the atmosphere and thus would be expected to show great constancy in concentration all around the globe. This is the case and the results of measurements are illustrated in Fig. 2.3.

Box 2.4

Chemical equilibrium

Many chemical reactions occur in both directions such that the products are able to re-form the reactants. For instance, in rainfall chemistry, we account for the hydrolysis (i.e. reaction with water) of aqueous formaldehyde to methylene glycol according to the equation:

$$HCHO_{(aq)} + H_2O_{(l)} \rightarrow H_2C(OH)_{2(aq)} \qquad \text{eq. 1}$$

(formaldehyde)   (methylene glycol)

but the reverse reaction also occurs:

$$H_2C(OH)_{2(aq)} \rightarrow HCHO_{(aq)} + H_2O_{(l)} \qquad \text{eq. 2}$$

such that the system is maintained in dynamic equilibrium, symbolised by:

$$HCHO_{(aq)} + H_2O_{(l)} \rightleftharpoons H_2C(OH)_{2(aq)} \qquad \text{eq. 3}$$

The relationship between the species at equilibrium is described in terms of the equation:

$$K = \frac{cH_2C(OH)_{2(aq)}}{cHCHO_{(aq)}.cH_2O} \qquad \text{eq. 4}$$

where c denotes the concentrations of the entities involved in the reaction. We should write this equation in terms of activities of the chemical species, which are the formal thermodynamic representations of concentration. However, in dilute solutions activity and concentration are almost identical. Dilute solutions, such as rainwater, are almost pure water. The activity of pure substances is defined as unity, so in the case of rainwater the equation can be simplified:

$$K = \frac{cH_2C(OH)_{2(aq)}}{cHCHO_{(aq)}} \qquad \text{eq. 5}$$

K is known as the equilibrium constant and in this case it has the value 2000. An equilibrium constant greater than unity suggests that equilibrium lies to the right-hand side and the forward reaction is favoured. Equilibrium constants vary with temperature, but not with concentration if the concentrations have been correctly expressed in terms of activities.

The equilibrium relationship is often called the law of mass action and may be remembered by the fact that an equilibrium constant is the product of the concentration of the products of a reaction divided by the product of the reactants, such that in general terms:

$$kA + lB \rightleftharpoons mC + nD \quad K \qquad \text{eq. 6}$$

$$k = \frac{cC^m.cD^n}{cA^k.cB^l} \qquad \text{eq.7}$$

It may be easier to grasp the notion of shifts in equilibrium in a qualitative way using the *Le Chatelier Principle*. This states that, if a system at equilibrium is perturbed, the system will react in such a way as to minimise this imposed change. Thus, looking at the formaldehyde equilibrium (eq. 3), any increase in HCHO in solution would be lessened by the tendency of the reaction to shift to the right, producing more methylene glycol.

2.4 Natural sources

Since the atmosphere can be treated, on a large scale, as if it were in steady state, we have a model that views the atmosphere as having sources, a reservoir (i.e. the atmosphere itself) and removal processes, all in delicate balance. The sources need to be quite stable over the long term. If they are not, then the balance will shift. In terms of our earlier analogy, the level in the leaking bucket will change.

The best-known, and most worrying, example of such a shift is the increasing magnitude of the CO_2 source because of the consumption of vast amounts of fossil fuel by human activities. This has given rise to a continuing increase in the CO_2 concentration in the atmosphere. The predicted rise in temperature, due to the greenhouse effect, is explored in more detail in Chapter 5.

There are many sources of trace components in the atmosphere, which can be divided into different categories, such as geochemical, biological and human or anthropogenic sources. Some of these sources are hard to categorise. Is a forest fire a geochemical, biological or human source (particularly if the forest was planted or the fire started through human activities)? Although our definitions can become a little blurred, it is nevertheless useful to categorise sources in this way.

2.4.1 Geochemical sources

Perhaps the largest geochemical sources are wind-blown dusts and sea sprays, which put huge amounts of solid material into the atmosphere. The dust is largely soil from arid regions of the Earth. If this dust is fine enough, it can spread over large areas of the globe and is important in redistributing material. Often, however, the chemical effects of the dust in the atmosphere are not particularly evident, because dusts are not chemically very reactive. In contrast, wind-blown sea spray places a more reactive entity into the atmosphere as salt particles.

The salt particles from the oceans are hygroscopic and under humid conditions these tiny NaCl crystals attract water and form a concentrated solution droplet or aerosol. Ultimately, this process can take part in cloud formation. The droplets can also be a site for important chemical reactions in the atmosphere. If strong acids (Box 2.5) in the atmosphere, perhaps nitric acid (HNO_3) or sulphuric acid (H_2SO_4), dissolve in these small droplets, hydrogen chloride (HCl) can be formed. It is thought that this process is an important source of HCl in the atmosphere:

$$H_2SO_{4\text{(in aerosol)}} + NaCl_{\text{(in aerosol)}} \rightarrow HCl_{\text{(g)}} + NaHSO_{4\text{(in aerosol)}} \qquad \text{eq. 2.6}$$

Incoming meteors also inject particles into the atmosphere. This is a very small source compared with wind-blown dust or forest fires, but meteors make their contribution to the upper parts of the atmosphere where the gas is at a low density. Here, a small contribution can be particularly significant and the metals ablated from incoming meteors enter a series of chemical reactions.

Volcanoes are an even larger souce of dust and particularly powerful eruptions can push dust into the stratosphere. It has long been known that volcanic particles can change global temperature by blocking out sunlight. They can also perturb the

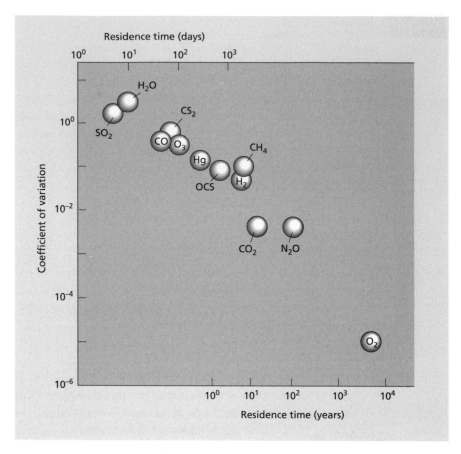

Fig. 2.3 Variability of trace and other components in the atmosphere as a function of residence time. Large coefficients of variation indicate higher variability. From Brimblecombe (1986).

chemistry at high altitudes. Along with the dust, volcanoes are huge sources of gases such as sulphur dioxide (SO_2), CO_2, HCl and hydrogen fluoride (HF). These gases can react in the stratosphere to provide a further source of particles, with H_2SO_4 being the most important particle produced indirectly from volcanoes.

It is important to realise that the volcanic source is a very discontinuous one, both in time and space. Large volcanic eruptions are infrequent. It may be that years pass without any really major eruptions and then suddenly more material is released in a single event than for many years previously. The eruptions occur in very specific locations where there are active volcanoes. In addition to massive eruptions that push great quantities of material into the upper parts of the atmosphere, we must not neglect smaller fumarolic emissions, from volcanic cracks and fissures, which gently release gases to the lower atmosphere over very long periods of time. The balance between these two volcanic sources is not accurately known, although for SO_2 it is probably about 50 : 50.

Radioactive elements in rocks (Box 2.6), most importantly potassium (K) and heavy elements such as radium (Ra), uranium (U) and thorium (Th), can release

Box 2.5

Acids and bases

Acids and bases are an important class of chemical compounds, because they exert special control over reactions in water. Traditionally acids have been seen as compounds that dissociate to yield hydrogen ions in water.

$$HCl_{(aq)} \rightarrow H^+_{(aq)} + Cl^-_{(aq)} \qquad \text{eq. 1}$$

The definition of an acid has, however, been extended to cover a wider range of substances by considering electron transfer. For example, boric acid (H_3BO_3), which helps control the acidity* of seawater, gains electrons from the hydroxide (OH^-) ion.

$$H_3BO_{3(aq)} + OH^-_{(aq)} \rightarrow B(OH)^-_{4(aq)} \qquad \text{eq. 2}$$

For most applications the simple definition is sufficient, and we might think of bases (or alkalis) as those substances which yield OH^- in aqueous solution.

$$NaOH_{(aq)} \rightarrow Na^+_{(aq)} + OH^-_{(aq)} \qquad \text{eq. 3}$$

Acids and bases react to neutralise each other, producing a dissolved salt plus water.

$$HCl_{(aq)} + NaOH_{(aq)} \rightarrow Cl^-_{(aq)} + Na^+_{(aq)} + H_2O_{(l)} \qquad \text{eq. 4}$$

Two classes of acids and bases are recognised — strong and weak. Hydrochloric acid (HCl) and sodium hydroxide (NaOH) (eqs 1 and 4) are treated as if they dissociate completely in solution to form ions, so they are termed 'strong'. Weak acids and bases dissociate only partly.

$$HCOOH_{(aq)} \rightleftharpoons H^+_{(aq)} + HCOO^-_{(aq)} \qquad \text{eq. 5}$$
(formic acid)

$$NH_4OH_{(aq)} \rightleftharpoons NH^+_{4(aq)} + OH^-_{(aq)} \qquad \text{eq. 6}$$
(ammonium hydroxide)

Dissociation is an equilibrium process and is conveniently described in terms of equilibrium constants for the acid (K_a) and alkaline (K_b) dissociation:

$$K_a = \frac{cH^+.cHCOO^-}{cHCOOH} = 1.77 \times 10^{-1} \text{ mol l}^{-1} \qquad \text{eq. 7}$$

$$K_b = \frac{cNH_4^+.cOH^-}{cNH_4OH} = 1.8 \times 10^{-5} \text{ mol l}^{-1} \qquad \text{eq. 8}$$

* The acidity of the oceans is usually defined by its pH, which is discussed in Box 2.10.

gases. Argon (Ar) arises from potassium decay and radon (Rn, a radioactive gas that has a half-life of 3.8 days) from radium decay. The uranium–thorium decay series results in the production of α particles, which are helium nuclei. Once these nuclei capture electrons, helium has effectively been added to the atmosphere.

Helium has not accumulated in the atmosphere over time because it is light enough to escape into space. The concentration of helium has been thus maintained in steady state through a balance of radioactive emanation from the crust and loss from the top of the atmosphere.

2.4.2 Biological sources

Unlike the geological sources, biology does not appear to be a large direct source of particles to the atmosphere, unless we consider forest fires a biological source. Table 2.2 shows that forest fires are quite an important source of carbon (C), i.e. soot particles.

The living forest also plays an important role in exchanging gases with the atmosphere. The major gases O_2 and CO_2 are, of course, involved in respiration and photosynthesis. However, forests also emit enormous quantities of trace organic compounds. Terpenes, such as pinene and limonene, give forests their wonderful odour. Forests are also important sources of organic acids, aldehydes and other organic compounds (Box 2.7).

Although forests are obvious as sources of gas it is the microorganisms that are especially important in generating atmospheric trace gases. Methane, which we have already discussed, is generated by reactions in anaerobic systems. Damp soils, as found in marshes or rice paddies, are important microbiologically dominated environments, as are the digestive tracts of ruminants such as cattle.

The soils of the Earth are rich in nitrogen compounds, giving rise to a whole range of active nitrogen chemistry that generates many nitrogenous trace gases. We can consider urea (NH_2CONH_2), present in animal urine, as a typical biologically generated nitrogen compound in the soil. Hydrolysis converts NH_2CONH_2 to ammonia (NH_3) and CO_2 according to the equation:

$$NH_2CONH_{2(aq)} + H_2O_{(l)} \rightarrow 2NH_{3(g)} + CO_{2(g)} \qquad \text{eq. 2.7}$$

If the soil where this hydrolysis occurs is alkaline (see Box 2.5), gaseous NH_3 can be released, although in acidic conditions it will react to form the ammonium ion (NH_4^+):

$$NH_{3(g)} + H^+_{(aq)} \rightarrow NH^+_{4(aq)} \qquad \text{eq. 2.8}$$

Table 2.2 Sources for particulate material in the atmosphere. From Brimblecombe (1986)

Source	Global flux (Tg year^{-1})
Forest fires	35
Dust	750
Sea salt	1500
Volcanic dust	50
Meteoritic dust	1

Box 2.6

Radioactive emanation

Some naturally occurring elements are radioactive. This means that their nuclei are unstable and spontaneously decay transforming the nucleus into that of a different element. Radioactive decay is written in equations that look a little like those for chemical reactions, but they need to express the atomic mass of the elements involved and the kinds of particles that are emitted in the form of radiation. The decay of potassium (^{40}K) can be written:

$$^{40}K \rightarrow {}^{40}Ar + \gamma \qquad \text{eq. 1}$$

In this transformation an electron of the potassium is captured by the nucleus and a proton within it is converted to a neutron. Excess energy is lost as a γ particle, which is essentially a photon that carries a large amount of electromagnetic energy. The important point for the atmosphere is the production of the stable form (isotope) of argon which emanates from the potassium-containing rocks of the earth and accumulates in the atmosphere. Some elements decay to produce radiation as an α particle, which is in fact a helium (He) nucleus.

$$^{238}U \rightarrow {}^{234}Th + \alpha \qquad \text{eq. 2}$$

As the α particle loses energy, it picks up electrons and eventually becomes ^4He in the atmosphere. Another source of helium is the decay of radium (Ra).

$$^{226}Ra \rightarrow {}^{222}Rn + \alpha \qquad \text{eq. 3}$$

which also produces the inert, but radioactive, gas radon (Rn).

Plants can absorb soil NH_3 or NH_4^+ directly and some microorganisms, such as *Nitrosomonas*, oxidise NH_3, using it as an energy source for respiration, in the same way that other cells use reduced carbon compounds. One possible reaction would be:

$$2NH_{3(g)} + 2O_{2(g)} \rightarrow N_2O_{(g)} + 3H_2O_{(g)} \qquad \text{eq. 2.9}$$

Here we can see a biological source for nitrous oxide (N_2O), an important and rather stable trace gas in the troposphere. In nature there are many other reactions of nitrogen compounds in soils that produce the gases: NH_3, N_2, N_2O and nitric oxide (NO).

Microorganisms in the oceans also prove to be an enormous source of atmospheric trace gases. Seawater is rich in dissolved sulphate and chloride (and to a lesser extent salts of the other halogens: fluorine (F), bromine (Br) and iodine (I)). Marine microorganisms metabolise these elements, for reasons that are not properly understood, to generate sulphur (S)- and halogen-containing trace gases. However, the nitrate concentration of surface seawater is so low that the oceans are effectively a nitrogen desert. This means that seawater is not such a large source of nitrogen-containing trace gases.

Box 2.7

Organic molecular structure

Organic molecules contain carbon, hydrogen and often some other non-metallic elements such as oxygen, nitrogen, sulphur or halogens such as chlorine. The complexity of organic molecules is such that it is often necessary to draw them as a simple picture rather than write the formula. As already seen in Box 2.2, bonds are written as lines between the molecules. We can see this principle used to show the configuration of some organic molecules discussed in this chapter.

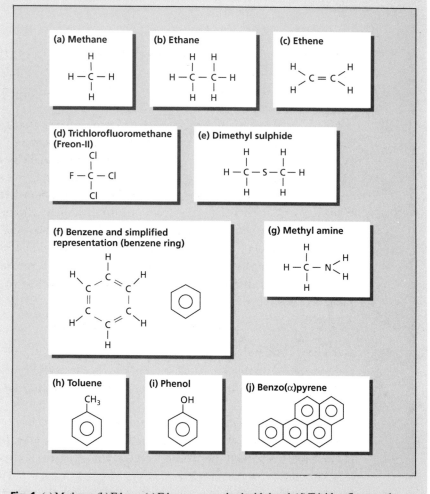

Fig. 1 (a) Methane. (b) Ethane. (c) Ethene — note the double bond. (d) Trichlorofluoromethane (Freon-11). (e) Dimethyl sulphide. (f) Benzene — the double bonds become delocalised so are symbolised by a circle. It is conventional to omit the H atoms from pictorial representations of benzene. (g) Methyl amine — containing an amine group (NH_2). (h) Toluene. (i) Phenol. (j) Benzo(α)pyrene — a polycyclic aromatic hydrocarbon (PAH). The benzene ring is particularly stable, which enables it to be the building block of larger molecules such as this one.

Organosulphides produced by marine microorganisms make a particularly significant contribution to the atmospheric sulphur burden. The most characteristic compound is dimethyl sulphide (DMS; $(CH_3)_2S$). This volatile compound is produced by marine phytoplankton, such as *Phaeocystis pouchetii*, in the upper layers of the ocean by the hydrolysis of beta-dimethylsulphoniopropionate (DMSP; $(CH_3)_2S^+CH_2CH_2COO^-$) to DMS and acrylic acid ($CH_2CHCOOH$):

$$(CH_3)_2S^+CH_2CH_2COO^-_{(aq)} \rightarrow (CH_3)_2S_{(g)} + CH_2CHCOOH_{(aq)} \qquad \text{eq. 2.10}$$

Another important sulphur compound released from the oceans is carbonyl sulphide (OCS). This can be produced by reaction between carbon disulphide (CS_2) and water:

$$CS_{2(g)} + H_2O_{(g)} \rightarrow OCS_{(g)} + H_2S_{(g)} \qquad \text{eq. 2.11}$$

and, although the flux to the atmosphere is smaller than that of DMS, its stability means that it will accumulate to higher concentrations. These sulphur gases have low solubility in water (Box 2.8), making them able easily to escape from the oceans into the atmosphere.

Halogenated organic compounds are well known in the atmosphere. Although these have an obvious human source, being present in cleaning fluids, fire extinguishers and aerosol propellants, there are also a wide range of biological sources. Methyl chloride (CH_3Cl) is the most abundant halocarbon in the atmosphere and arises primarily from poorly understood marine sources, although terrestrial microbiological processes and biomass burning also contribute. Bromine- and iodine-containing organic compounds are also released from the oceans and the distribution of this marine iodine over land masses represents an important source of this essential trace element for mammals. As one might predict, the iodine-deficiency disease, goitre, has been common in regions remote from the oceans.

2.5 Reactivity of trace substances in the atmosphere

Gases with short residence times in the atmosphere are clearly those which can be easily removed. Some may be removed by being absorbed by plants or solids or into water. However, chemical reactions are the usual reason for a gas having a short residence time.

What makes gases react in the atmosphere? It turns out that most of the trace gases listed in Table 2.3 are not very reactive with the major components of air. In fact, the most important reactive entity in the atmosphere is a fragment of a water molecule, the hydroxyl (OH) radical. This radical (a reactive molecular fragment) is formed by the photochemically initiated reaction sequence, started by the photon of light, $h\nu$:

$$O_{3(g)} + h\nu \rightarrow O_{2(g)} + O_{(g)} \qquad \text{eq. 2.12}$$

$$O_{(g)} + H_2O_{(g)} \rightarrow 2OH_{(g)} \qquad \text{eq. 2.13}$$

Box 2.8

Gas solubility

The solubility of gases in liquids is often treated as an equilibrium process. Take the dissolution of carbonyl sulphide (OCS) as an example.

$$OCS_{(g)} \rightleftharpoons OCS_{(aq)} \qquad \text{eq. 1}$$

where $OCS_{(g)}$ and $OCS_{(aq)}$ represent the concentration of the substance carbonyl sulphide in the gas and liquid phase. This equilibrium relationship is often called Henry's law, after the English physical chemist who worked c. 1800. The Henry's law constant (K_H) describes the equilibrium. Using pressure (p) to describe the concentration (c) of $OCS_{(g)}$ in the gas phase we have:

$$K_H = \frac{cOCS_{(aq)}}{pOCS_{(g)}} \qquad \text{eq. 2}$$

If we take the atmosphere as the unit of pressure and mol l^{-1} as the unit of concentration, the Henry's law constant will have the units mol l^{-1} atm^{-1}. The larger the values of this constant, the more soluble the gas. Table 1 shows that a gas like hydrogen peroxide is very soluble, oxygen very much less so.

Many quite important gases have only limited solubility, but often they can react in water, which enhances their solubility. Take the simple dissolution of formaldehyde (HCHO), which readily hydrolyses to methylene glycol ($H_2C(OH)_2$).

$$HCHO_{(g)} \rightleftharpoons HCHO_{(aq)} \qquad \text{eq. 3}$$

$$HCHO_{(aq)} + H_2O_{(l)} \rightleftharpoons H_2C(OH)_{2(aq)} \qquad \text{eq. 4}$$

The second equilibrium lies so far to the right that solubility is enhanced by a factor of about 2000.

Table 1 Some Henry's law constants at 15°C

Gas	K_H (mol l^{-1} atm^{-1})
Hydrogen peroxide	2×10^5
Ammonia	90
Formaldehyde	1.7
Dimethyl sulphide	0.14
Carbonyl sulphide	0.035
Ozone	0.02
Oxygen	0.0015
Carbon monoxide	0.001

The OH radical can react with many compounds in the atmosphere and thus it has a short residence time. The rates are faster than with abundant gases such as O_2.

The reaction between nitrogen dioxide (NO_2) and the OH radical leads to the formation of HNO_3 an important contributor to acid rain.

$$NO_{2(g)} + OH_{(g)} \rightarrow HNO_{3(g)} \qquad\qquad \text{eq. 2.14}$$

In contrast, kinetic measurements in the laboratory (which aim at determining the speed of reaction) show that gases that have slow rates of reaction with the OH radical have a long residence time in the atmosphere. Table 2.3 shows that OCS, N_2O and even CH_4 have long residence times. The CFCs (chlorofluorocarbons: refrigerants and aerosol propellants) also have very limited reactivity with OH. Gases like these build up in the atmosphere and eventually leak across the tropopause into the stratosphere. Here there is a very different chemistry taking place, no longer dominated by OH but by reactions which involve atomic oxygen (i.e. O). Gases that react with atomic oxygen in the stratosphere can interfere with the production of O_3:

$$O_{(g)} + O_{2(g)} \rightarrow O_{3(g)} \qquad\qquad \text{eq. 2.15}$$

and can be responsible for the depletion of the stratospheric O_3 layer (Box 2.9).

This means that CFCs are prime candidates for causing damage to stratospheric O_3 (a topic discussed further in Chapter 5). We should note that nitrogen compounds are also damaging to O_3 if they can be transported to the stratosphere, because they are involved in similar reaction sequences. We have already seen that tropospheric NO_2 is unlikely to be transferred into the stratosphere (eq. 2.14). It was, however, nitrogen compounds from the exhausts of commercial supersonic aircraft flying at high altitude that were the earliest suggested contaminants of

Table 2.3 Naturally occurring trace gases of the atmosphere. From Brimblecombe (1986)

	Residence time	Concentration (ppb)
Carbon dioxide	4 years	360 000
Carbon monoxide	0.1 year	100
Methane	3.6 years	1 600
Formic acid	10 days	1
Nitrous oxide	20–30 years	300
Nitric oxide	4 days	0.1
Nitrogen dioxide	4 days	0.3
Ammonia	2 days	1
Sulphur dioxide	3–7 days	0.01–0.1
Hydrogen sulphide	1 day	0.05
Carbon disulphide	40 days	0.02
Carbonyl sulphide	1 year	0.5
Dimethyl sulphide	1 day	0.001
Methyl chloride	30 days	0.7
Methyl iodide	5 days	0.002
Hydrogen chloride	4 days	0.001

concern. In this case the gases did not have to be unreactive and slowly transfer to the stratosphere, but were directly injected from aircraft engines. A large stratospheric transport fleet never came about, so attention has now turned to N_2O, a much more inert oxide of nitrogen produced at ground level and quite capable of getting into the stratosphere. This gas is produced both from biological activities in fertile soils (Section 2.4.2) and by a range of combustion processes — most interestingly, automobile engines with catalytic converters.

Finally, we should note that some reactions lead to the formation of particles in the atmosphere. Most particles are effectively removed by rainfall and thus have residence times close to the 4–10 days of atmospheric water. In contrast, very small particles in the 0.1–1 μm size range are not very effectively removed by rain droplets and have rather longer residence times.

2.6 The urban atmosphere

In the section above we began to look at human influence on the atmosphere. The changes wrought by humans are important, though often subtle on the global scale. It is in the urban atmosphere where human influence shows its clearest impact, so it is necessary to treat the chemistry of the urban atmosphere as a special case.

In urban environments there are pollutant compounds emitted to the atmosphere directly and these are called primary pollutants. Smoke is the archetypical example of a primary pollutant. However, many compounds undergo reactions in the atmosphere, as we have seen in the section above. The products of such

---

Box 2.9

Ozone

The formation of ozone (O_3) is a photochemical process which uses the energy involved in light. The shorter the wavelength of light, the larger the amount of energy it carries. It requires ultraviolet (UV) radiation of wavelength less than 242 nm to have sufficient energy to split the oxygen molecule (O_2) apart.

$$O_{2(g)} + h\nu \rightarrow O_{(g)} + O_{(g)} \qquad \text{eq. 1}$$

The UV photon here is symbolised by $h\nu$. Once oxygen atoms (O) have been formed, they can react with O_2.

$$O_{2(g)} + O_{(g)} \rightarrow O_{3(g)} \qquad \text{eq. 2}$$

The production of O_3 by this photochemical process can be balanced against the reactions that destroy O_3. These may be written:

$$O_{3(g)} + h\nu \rightarrow O_{2(g)} + O_{(g)} \qquad \text{eq. 3}$$

$$O_{3(g)} + O_{(g)} \rightarrow 2O_{2(g)} \qquad \text{eq. 4}$$

together with an additional reaction describing the destruction process for oxygen atoms:

**Box 2.9
Cont.**

$$O_{(g)} + O_{(g)} + M_{(g)} \rightarrow O_{2(g)} + M_{(g)} \qquad \text{eq. 5}$$

Note the presence of the 'third body' M, which carries away excess energy during the reaction. The third body would typically be O_2 or a nitrogen molecule (N_2). Without this 'third body' the O_2 that formed might split apart again. These principal 'oxygen-only' reactions do not fully describe the O_3 chemistry and we need to consider reactions involving hydrogen (H)-containing, nitrogen (N)-containing, and chlorine (Cl)-containing species.

$$OH_{(g)} + O_{3(g)} \rightarrow O_{2(g)} + HO_{2(g)} \qquad \text{eq. 6}$$

$$HO_{2(g)} + O_{(g)} \rightarrow OH_{(g)} + O_{2(g)} \qquad \text{eq. 7}$$

which sum:

$$O_{3(g)} + O_{(g)} \rightarrow 2O_{2(g)} \qquad \text{eq. 8}$$

Similar reactions can be written for other species, e.g. nitric oxide (NO), which arises from supersonic aircraft, or nitrous oxide (N_2O), which crosses the tropopause into the stratosphere:

$$NO_{(g)} + O_{3(g)} \rightarrow O_{2(g)} + NO_{2(g)} \qquad \text{eq. 9}$$

$$NO_{2(g)} + O_{(g)} \rightarrow NO_{(g)} + O_{2(g)} \qquad \text{eq. 10}$$

and $N_2O_{(g)}$ can enter reaction 9 via the initial step:

$$N_2O_{(g)} + O_{(g)} \rightarrow 2NO_{(g)} \qquad \text{eq. 11}$$

The sequence for chlorine from chlorofluorocarbons (CFCs) is:

$$O_{3(g)} + Cl_{(g)} \rightarrow O_{2(g)} + ClO_{(g)} \qquad \text{eq. 12}$$

$$ClO_{(g)} + O_{(g)} \rightarrow O_{2(g)} + Cl_{(g)} \qquad \text{eq. 13}$$

All these three reaction pairs (eqs 7–8, 9–10, 12–13) sum in such a way as to destroy O_3 and atomic oxygen while restoring the OH, NO or Cl molecules. They can thus be regarded as catalysts for O_3 destruction. Catalysts are chemical species which facilitate a reaction, but undergo no net consumption or production in a reaction. The important point of these catalytic reaction chains, in the chemistry of stratospheric O_3, is that a single pollutant molecule can be responsible for the destruction of a large number of O_3 molecules.

reactions are called secondary pollutants. Thus, many primary pollutants can react to produce secondary pollutants. It is the distinction between primary and secondary pollution that now governs our understanding of the difference between two quite distinct types of air pollution that affect major cities of the world.

2.6.1   London smog — primary pollution

Urban pollution is largely the product of combustion processes. In ancient times cities such as Imperial Rome experienced pollution problems due to wood smoke. However, it was the transition to fossil fuel burning that caused the rapid development of air pollution problems. The inhabitants of London have burnt coal since the thirteenth century. Concern and attempts to regulate coal burning began almost immediately, as there was a perceptible and rather strange smell associated with it. Medieval Londoners thought this smell might be associated with disease.

Fuels usually consist of hydrocarbons, except in particularly exotic applications such as rocketry, where nitrogen, aluminium (Al) and even beryllium (Be) are sometimes used. We can describe normal fuel combustion according to the equation:

$$\text{`4CH'} + 5O_{2(g)} \rightarrow 4CO_{2(g)} + 2H_2O_{(g)} \qquad \text{eq. 2.16}$$
$$\text{fuel} + \text{oxygen} \rightarrow \text{carbon dioxide} + \text{water}$$

This would not seem an especially dangerous activity as neither CO_2 nor water is particularly toxic. However, let us consider a situation where there is not enough O_2 during combustion, i.e. as might occur inside an engine or boiler. The equation might now be written:

$$\text{`4CH'} + 3O_{2(g)} \rightarrow 4CO_{(g)} + 2H_2O_{(g)} \qquad \text{eq. 2.17}$$
$$\text{coal} + \text{oxygen} \rightarrow \text{carbon monoxide} + \text{water}$$

Here we have produced carbon monoxide (CO), a poisonous gas. With even less oxygen we can get carbon (i.e. smoke):

$$\text{`4CH'} + O_{2(g)} \rightarrow 4C_{(s)} + 2H_2O_{(g)} \qquad \text{eq. 2.18}$$
$$\text{coal} + \text{oxygen} \rightarrow \text{`smoke'} + \text{water}$$

At low temperatures, in situations where there is relatively little O_2, pyrolysis reactions (i.e. reactions where decomposition takes place as a result of heat) may cause a rearrangement of atoms that can lead to the formation of polycyclic aromatic hydrocarbons during combustion (see Box 2.7). The most notorious of these is benzo(α)pyrene, B(α)P, a cancer-inducing compound.

Thus, although the combustion of fuels would initially seem a harmless activity, it can produce a range of pollutant carbon compounds. When the earliest steam engines were being designed, engineers saw that an excess of oxygen would help convert all the carbon to CO_2. To overcome this they adopted a philosophy of 'burning your own smoke', even though this required considerable skill to implement and was consequently of only limited success.

In addition to these problems, contaminants within the fuel can also cause air pollution. The most common and worrisome impurity in fossil fuels is sulphur (S), partly present as the mineral pyrite, FeS_2. There may be as much as 6% sulphur in some coals and this is converted to SO_2 on combustion:

$$4FeS_{2(s)} + 11O_{2(g)} \rightarrow 8SO_{2(g)} + 2Fe_2O_3 \qquad \text{eq. 2.19}$$

There are other impurities in fuels too, but sulphur has always been seen as most characteristic of the air pollution problems of cities.

If we look at the composition of various fuels (Table 2.4), we see that they contain quite variable amounts of sulphur. The highest amounts of sulphur are found in coals and in fuel oils. These are the fuels used in stationary sources such as boilers, furnaces (and traditionally steam engines), domestic chimneys, steam turbines and power stations. Thus, the main source of sulphur pollution, and indeed smoke, in the urban atmosphere is the stationary source. Smoke too is mainly associated with stationary sources. Steam trains and boats caused the occasional problem, but it was the stationary source that was most significant.

For many people, SO_2 and smoke came to epitomise the traditional air pollution problems of cities. Smoke and SO_2 are obviously primary pollutants because they are formed directly at a clearly evident pollutant source and enter the atmosphere in that form.

Classical air pollution incidents in London occurred under damp and foggy conditions in the winter. Fuel use was at its highest and the air near-stagnant. The presence of smoke and fog together led to the invention of the word smog (sm[oke and f]og), now often used to describe air pollution in general (Fig. 2.4). Sulphur dioxide is fairly soluble so could dissolve into the water that condensed around smoke particles.

$$SO_{2(g)} + H_2O_{(l)} \rightleftharpoons H^+_{(aq)} + HSO^-_{3(aq)} \qquad \text{eq. 2.20}$$

Traces of metal contaminants (iron (Fe) or manganese (Mn)) catalysed the conversion of dissolved SO_2 to H_2SO_4 (see box 2.9 for a definition of catalyst).

$$2HSO^-_{3(aq)} + O_{2(aq)} \rightleftharpoons 2H^+_{(aq)} + 2SO^{2-}_{4(aq)} \qquad \text{eq. 2.21}$$

Sulphuric acid has a great affinity for water so the droplet absorbed more water. Gradually the droplets grew and the fog thickened, attaining very low pH values (Box 2.10).

Table 2.4 Sulphur content of fuels

Fuel	S (percentage by weight)
Coal	0.2–7.0
Fuel oils	0.5–4.0
Coke	1.5–2.5
Diesel fuel	0.3–0.9
Petrol	0.1
Kerosene	0.1
Wood	Very small
Natural gas	Very small

Fig. 2.4 The London smog of 1952. Photograph courtesy of Popperfoto Northampton, UK.

Terrible fogs plagued London at the turn of the last century when Sherlock Holmes and Jack the Ripper paced the streets of the metropolis. The incidence of bronchial disease invariably rose at times of prolonged winter fog — little wonder, considering that the fog droplets contained H_2SO_4. Medical registrars in Victorian England realised that the fogs were affecting health, but they, along with others, were not able to legislate smoke out of existence. Even where there was a will, and indeed there were enthusiasts in both Europe and North America who strove for change, the technology was far too naïve to achieve really noticeable improvements. The improvements that did come about were often due to changes in fuel, in location of industry or in climate.

2.6.2   Los Angeles smog — secondary pollution

The air pollutants that we have been discussing so far have come from stationary sources. Traditionally, industrial and domestic activities in large cities burnt coal. The transition to petroleum-derived fuels this century has seen the emergence of an entirely new kind of air pollution. This newer form of pollution is the result of the greater volatility of liquid fuels. The motor vehicle is such an important consumer of liquid fuels that it has become a major source of contemporary air pollution. However, the pollutants really responsible for causing the problems are not themselves emitted by motor vehicles. Rather, they form in the atmosphere. These

Box 2.10

The pH scale

The acidity of aqueous solutions is frequently described in terms of the pH scale. Acids (see Box 2.5) give rise to hydrogen ions (H⁺) in solution and the pH value of such a solution is defined:

$$pH = -\log_{10}(cH^+_{(aq)})$$ eq. 1

Strictly speaking $cH^+_{(aq)}$ should mean hydrogen ion activity ($aH^+_{(aq)}$), but at the moment we can imagine that as being similar to the concentration in units of mol l⁻¹.

We can write a similar relationship identifying pOH.

$$pOH = -\log_{10}(cOH^-_{(aq)})$$ eq. 2

However, pH is related to pOH through the equilibrium describing the dissociation of water:

$$H_2O \rightleftharpoons H^+ + OH^-, \text{ i.e. } K_w = 10^{-14} = cH^+.cOH^-$$ eq. 3

such that pH = 14 – pOH.

It is important to notice that this is a logarithmic scale, so it is not appropriate to average pH values of solutions (although one can average H⁺ concentrations).

On the pH scale, 7 is regarded as neutral. This is the point where $cH^+ = cOH^-$. There are a number of other important values on the scale (conventionally made to stretch from 0 to 14) that are relevant to the environment.

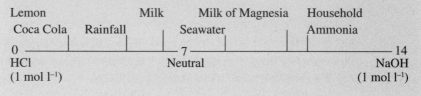

secondary pollutants are formed from the reactions of primary pollutants, such as NO and unburnt fuel, which come directly from the automobiles. Chemical reactions that produce the secondary pollutants proceed most effectively in sunlight, so the resulting air pollution is called photochemical smog.

Photochemical smog was first noticed in Los Angeles during the Second World War. Initially it was assumed to be similar to the air pollution that had been experienced elsewhere, but conventional smoke abatement techniques failed to lead to any improvement. In the 1950s it became clear that this pollution was different, and the experts were baffled. A. Haagen-Smit, a biochemist studying vegetation damage in the Los Angeles basin, realised that the smog was caused by reactions of automobile exhaust vapours in sunlight.

Although air pollution and smoke have traditionally been closely linked, there were always those who thought there was more to air pollution than just smoke. We can now see how impurities in fuel give rise to further pollutants. In addition, the fact that we burn fuels, not in O_2, but in air has important consequences. We have learnt that air is a mixture of O_2 and N_2. At high temperature, in a flame, molecules in air may fragment, and even the relatively inert N_2 molecule can undergo reaction:

$$O_{(g)} + N_{2(g)} \rightarrow NO_{(g)} + N_{(g)} \qquad \text{eq. 2.22}$$

$$N_{(g)} + O_{2(g)} \rightarrow NO_{(g)} + O_{(g)} \qquad \text{eq. 2.23}$$

Equation 2.23 produces an oxygen atom, which can re-enter eq. 2.22. Once an oxygen atom is formed in a flame, it will be regenerated and contribute to a whole chain of reactions that produce NO. If we add these two reactions we get:

$$N_{2(g)} + O_{2(g)} \rightarrow 2NO_{(g)} \qquad \text{eq. 2.24}$$

The equations show how nitrogen oxides are generated in flames. They arise because we burn fuels in air rather than just in O_2. In addition, some fuels contain nitrogen compounds as impurities, so the combustion products of these impurities are a further source of nitrogen oxides (i.e. NO_x, the sum of NO and NO_2).

Oxidation of nitric oxide in smog gives nitrogen dioxide (Box 2.11), which is a brown gas. This colour means that it absorbs light and is photochemically active and undergoes dissociation.

$$NO_{2(g)} + h\nu \rightarrow O_{(g)} + NO_{(g)} \qquad \text{eq. 2.25}$$

Equation 2.25 thus reforms the nitric oxide, but also gives an isolated and reactive oxygen atom, which can react to form O_3:

$$O_{(g)} + O_{2(g)} \rightarrow O_{3(g)} \qquad \text{eq. 2.26}$$

Ozone is the single pollutant that most clearly characterises photochemical smog. However, O_3, which we regard as such a problem, is not emitted by automobiles (or any major polluter). It is a secondary pollutant.

The volatile organic compounds released through the use of petroleum fuels serve to aid the conversion of NO to NO_2. The reactions are quite complicated, but we can simplify them by using a very simple organic molecule such as CH_4 to represent the petroleum vapour from vehicles:

$$CH_{4(g)} + 2O_{2(g)} + 2NO_{(g)} \xrightarrow{h\nu} H_2O_{(g)} + HCHO_{(g)} + 2NO_{2(g)} \qquad \text{eq. 2.27}$$

We can see two things taking place in this reaction. Firstly, the automobile hydrocarbon is oxidised to an aldehyde (i.e. a molecule with a CHO group). In the reaction above it is formaldehyde (HCHO). Aldehydes are eye irritants and, at high concentrations, also carcinogens. This equation simply shows the net reactions in photochemical smog. In Box 2.11 the process is given in more detail. In particular, it emphasises the role of the ubiquitous OH radical in promoting chemical reactions in the atmosphere.

The smog found in the Los Angeles basin (Plate 2.1, facing p. 142) is very different from that we have previously described as typical of coal-burning cities.

There is no fog when Los Angeles smog forms, and visibility does not decline to just a few metres, as was typical of London fogs. Of course, the Los Angeles smog forms best on sunny days. London fogs are blown away by wind, but the gentle sea breezes in the Los Angeles basin can hold the pollution in against the mountains and prevent it from escaping out to sea. The pollution cannot rise in the atmosphere because it is trapped by an inversion layer: the air at ground level is cooler than that aloft, so that a cap of warm air prevents the cooler air from rising and dispersing the pollutants. A fuller list of the differences between Los Angeles- and London-type smogs is given in Table 2.5.

2.7 **Air pollution and health**

We saw in Section 2.6.1 that the acid-laden smoke particles in the London atmosphere caused great harm to human health. Pollutants in the atmosphere still cause concern because of their effect on human health, although today we need to consider a wider range of potentially harmful trace substances. The photochemical smog encountered ever more widely in modern cities gives urban atmospheres that are unlike the smoky air of cities in the past. Petrol as a fuel, unlike coal, produces little smoke.

The two gases that particularly characterise photochemical smog, O_3 and nitrogen oxides, cause respiratory problems. Ozone impairs lung function, while nitrogen oxides, at high concentrations, are particularly likely to cause problems for asthmatics. Oxygen-containing compounds, such as aldehydes, cause eye, nose and throat irritation, as well as headaches, during periods of smog. Eye irritation is a frequent complaint in Los Angeles and other photochemically polluted cities. This eye irritation is particularly associated with a group of nitrogen-containing organic compounds. They are produced in reactions of nitrogen oxides with various organic compounds in the smog (see Box 2.11). The best known of these nitrogen-containing eye irritants is peroxyacetylnitrate, often called PAN.

Photochemical smog is not the only pollution problem created by vehicles. Automobiles are also associated with other pollutants such as lead (Pb) and benzene (C_6H_6). The success of lead tetralkyl compounds as antiknock agents for improving the performance of automotive engines has meant that, in countries with high car use, very large quantities of lead have been mobilised. This lead has been widely dispersed, but particularly large quantities have been deposited in cities and near heavily used roads. Lead is a toxin and has been linked with several environmental health problems. Perhaps the most worrying evidence has come from studies (although difficult to reproduce) which suggest a decline in intelligence among children exposed to quite low concentrations of lead.

Unleaded petrol was introduced in the USA in the 1970s so that catalytic converters could be used on cars. Since then, unleaded petrol has become used more widely. There is evidence that blood lead concentrations have dropped in parallel with the declining automotive source of lead. Nevertheless, the decrease in

Box 2.11

Reactions in photochemical smog

Reactions involving nitrogen oxides (NO and NO_2) and ozone (O_3) lie at the heart of photochemical smog.

$$NO_{2(g)} + h\nu \text{ (less than 310 nm)} \rightarrow O_{(g)} + NO_{(g)} \qquad \text{eq. 1}$$

$$O_{(g)} + O_{2(g)} + M_{(g)} \rightarrow O_{3(g)} + M_{(g)} \qquad \text{eq. 2}$$

$$O_{3(g)} + NO_{(g)} \rightarrow O_{2(g)} + NO_{2(g)} \qquad \text{eq. 3}$$

It is conventional to imagine these processes that destroy and produce nitrogen dioxide (NO_2) as in a kind of equilibrium, which is represented by a notional equilibrium constant relating the partial pressures of the two nitrogen oxides and O_3.

$$K = \frac{pNO \cdot pO_3}{pNO_2} \qquad \text{eq. 4}$$

If we were to increase NO_2 concentrations (in a way that did not use O_3), then the equilibrium could be maintained by increasing O_3 concentrations. This happens in the photochemical smog through the mediation of hydroxyl (OH) radicals in the oxidation of hydrocarbons. Here we will use methane (CH_4) as a simple example of the process.

$$OH_{(g)} + CH_{4(g)} \rightarrow H_2O_{(g)} + CH_{3(g)} \qquad \text{eq. 5}$$
(methane)

$$CH_{3(g)} + O_{2(g)} \rightarrow CH_3O_{2(g)} \qquad \text{eq. 6}$$

$$CH_3O_{2(g)} + NO_{(g)} \rightarrow CH_3O_{(g)} + NO_{2(g)} \qquad \text{eq. 7}$$

$$CH_3O_{(g)} + O_{2(g)} \rightarrow HCHO_{(g)} + HO_{2(g)} \qquad \text{eq. 8}$$
(formaldehyde)

$$HO_{2(g)} + NO_{(g)} \rightarrow NO_{2(g)} + OH_{(g)} \qquad \text{eq. 9}$$

These reactions represent the conversion of nitric oxide (NO) to NO_2 and a simple alkane such as CH_4 to an aldehyde, here formaldehyde (HCHO). Note that the OH radical is regenerated, so can be thought of as a kind of catalyst. Although the reaction will happen in photochemical smog, the attack of the OH radical is much faster on larger and more complex organic molecules. Aldehydes may also undergo attack by OH radicals:

$$CH_3CHO_{(g)} + OH_{(g)} \rightarrow CH_3CO_{(g)} + H_2O_{(g)} \qquad \text{eq. 10}$$
(acetaldehyde)

$$CH_3CO_{(g)} + O_{2(g)} \rightarrow CH_3COO_{2(g)} \qquad \text{eq. 11}$$

Box 2.11 Cont.

$$CH_3COO_{2(g)} + NO_{(g)} \rightarrow NO_{2(g)} + CH_3CO_{2(g)} \qquad \text{eq. 12}$$

$$CH_3CO_{2(g)} \rightarrow CH_{3(g)} + CO_{2(g)} \qquad \text{eq. 13}$$

The methyl radical (CH_3) in eq. 13 may re-enter at eq. 6.
An important branch to this set of reactions is:

$$CH_3COO_{2(g)} + NO_{2(g)} \rightarrow CH_3COO_2NO_{2(g)} \qquad \text{eq. 14}$$
$$\text{(PAN)}$$

leading to the formation of the eye irritant peroxyacetylnitrate (PAN)

atmospheric lead may not yet be enough to reduce possible subtle health effects in children to a satisfactory level. This is because children have a high intake of food relative to their body weight. Thus children are more likely than adults to consume a significant amount of their intake of lead with food and water. Although some of the lead in foodstuffs may have come from the atmosphere, lead in foods may also result from processing.

Table 2.5 Comparison of Los Angeles and London smog. From Raiswell *et al.* (1980)

Characteristic	Los Angeles	London
Air temperature	24 to 32 °C	−1 to 4 °C
Relative humidity	< 70%	85% (+ fog)
Type of temperature inversion	Subsidence, at 1000 m	Radiation (near ground) at a few-hundred metres
Wind speed	< 3 ms^{-1}	Calm
Visibility	< 0.8–1.6 km	< 30 m
Months of most frequent occurrence	Aug.–Sept.	Dec.–Jan.
Major fuels	Petroleum	Coal and petroleum products
Principal constituents	O_3, NO, NO_2, CO, organic matter	Particulate matter, CO, S compounds
Type of chemical reaction	Oxidative	Reductive
Time of maximum occurrence	Midday	Early morning
Principal health effects	Temporary eye irritation (PAN)	Bronchial irritation, coughing (SO_2/smoke)
Materials damaged	Rubber cracked (O_3)	Iron, concrete

Benzene is another pollutant component of automotive fuels. It occurs naturally in crude oil and is a useful component because it can prevent pre-ignition in unleaded petrol (the production process is usually adjusted so that the benzene concentration is about 5%). There is evidence that in some locations (e.g. Mexico City), where there has been a switch to fuels with high concentrations of aromatic hydrocarbons, there has been a sharp increase in photochemical smog. This is due to the high reactivity of these hydrocarbons in the urban atmosphere. This problem should draw our attention to the way in which the solution of one obvious environmental problem (lead from petrol) may introduce a second rather more subtle problem (i.e. increased photochemical smog from reactive aromatic compounds).

Benzene is also a potent carcinogen. It appears that more than 10% of the benzene used by society (33 Mtonne year^{-1}) is ultimately lost to the atmosphere. High concentrations of benzene can be found in the air of cities and these concentrations may increase the number of cancers. Exposure is complicated by the importance of other sources of benzene to humans, e.g. tobacco smoke. Toluene ($C_6H_5CH_3$) is another aromatic compound present in large concentrations in petrol. Toluene is less likely to be a carcinogen than benzene but it has some undesirable effects. Perhaps most importantly it reacts to form a PAN-type compound, peroxybenzoyl nitrate, which is a potent eye irritant.

2.8 Effects of air pollution

In the past, when smoke was the predominant air pollutant, its effects were easy to see. Even today, black incrustations on older buildings in many large cities are still evident. In addition, clothes were soiled, curtains and hangings were blackened and plant growth was affected. City gardeners carefully chose only the most resistant plants. A few decades ago, the trees around industrial centres became so blackened that light-coloured moths were no longer camouflaged. Melanic (dark) forms became more common because predators could see them less easily. Plants are also very sensitive to SO_2 and one of the first effects seems to be the inhibition of photosynthesis.

The traditional smog generated by coal burning contained SO_2 and its oxidation product, H_2SO_4, in addition to smoke. Sulphuric acid is a powerful corrosive agent and rusts iron bars and weathers building stones. Architects sometimes complained of layers of sulphate damage 10 cm thick on calcareous stone through the reaction:

$$H_2SO_{4(aq)} + CaCO_{3(s)} + H_2O_{(l)} \rightarrow CO_{2(g)} + CaSO_{4(s)}.2H_2O_{(l)} \qquad \text{eq. 2.28}$$

Sulphuric acid converts limestone ($CaCO_3$) into gypsum ($CaSO_4.2H_2O$). The deterioration is severe because gypsum is soluble and dissolves in rain. Perhaps more importantly, gypsum occupies a larger volume than limestone, which adds mechanical stress so that the stone almost explodes from within.

Diesel-powered vehicles are increasingly common in Europe, and this is not just confined to large vehicles. Today many cars are diesel-powered, taking advantage of potentially lower fuel costs. Diesel fuels have had the advantage of being unleaded. In spite of these advantages, the fuel injection process of the diesel engine leads to the fuel dispersing as droplets within the engine. These may not always burn completely, so diesel engines can produce large quantities of smoke if not properly maintained. Diesel smoke now makes a significant contribution to the soiling quality of urban air. In addition, the particles are rich in polyaromatic hydrocarbons (PAH), which are carcinogens.

In the modern urban atmosphere, O_3 may be the pollutant of particular concern for health. However, it is a reactive gas that will also attack the double bonds of organic molecules (see Box 2.7) very readily. Rubber is a polymeric material with many double bonds, so it is degraded and cracked by O_3. Tyres and windscreen wiper blades are especially vulnerable to oxidants, although newer synthetic rubbers have double bonds protected by other chemical groups, which can make them more resistant to damage by O_3.

Many pigments and dyes are also attacked by O_3. The usual result of this is that the dye fades. This means that it is important for art galleries in polluted cities to filter their air, especially where they house collections of paintings using traditional colouring materials, which are especially sensitive. Nitrogen oxides, associated with photochemical smogs, can also damage pigments. It is possible that nitrogen oxides may also increase the rate of damage to building stone, but it is not really clear how this takes place. Some have argued that NO_2 increases the efficiency of production of H_2SO_4 on stone surfaces in those cities that have moderate SO_2 concentrations.

$$SO_{2(g)} + NO_{2(g)} + H_2O_{(1)} \rightarrow NO_{(g)} + H_2SO_{4(aq)} \qquad \text{eq. 2.29}$$

Others have suggested that the nitrogen compounds in polluted atmospheres enable microorganisms to grow more effectively on stone surfaces and enhance the biologically mediated damage. There is also the possibility that gas-phase reactions produce HNO_3 (see cq. 2.14) and that this deposits directly on to calcareous stone.

Finally, we should remember that it is not just materials that are damaged by photochemical smog, since plants are especially sensitive to the modern atmospheric pollutants. Recollect that it was this sensitivity that led Haagen-Smit to recognise the novelty of the Los Angeles smog. Ozone damages plants by changing the 'leakiness' of cells to important ions such as potassium. Early symptoms of such injury appear as water-soaked areas on the leaves.

Urban air pollution remains an issue of much public concern. While it is true that in many cities the traditional problems of smoke and SO_2 from stationary sources are a thing of the past, new problems have emerged. In particular, the automobile and heavy use of volatile fuels have made photochemical smog a widespread occurrence. This has meant that there has been a parallel rise in legislation to lower the emission of these organic compounds to the atmosphere.

Box 2.12

Acidification of rain droplets

In Box 2.8 we saw the way reactions affect the solubility of gases. It is possible for some gases to undergo more complex hydration reactions in water, which influence its pH (see Box 2.10). The best known of these is the dissolution of carbon dioxide (CO_2), which gives natural rainwater its characteristic pH.

$$CO_{2(g)} + H_2O_1 \rightleftharpoons H_2CO_{3(aq)} \qquad \text{eq. 1}$$

$$H_2CO_{3(aq)} \rightleftharpoons H^+_{(aq)} + HCO^-_{3(aq)} \qquad \text{eq. 2}$$

$$HCO^-_{3(aq)} \rightleftharpoons H^+_{(aq)} + CO^{2-}_{3(aq)} \qquad \text{eq. 3}$$

Equation 3 is not important in the atmosphere, so the pH of a droplet of water in equilibrium with atmospheric CO_2 can be determined by combining the first two equilibrium constant equations that govern the dissolution (i.e. Henry's law, as discussed in Box 2.8) and dissociation. If carbonic acid (H_2CO_3) is the only source of protons, then cH^+ must necessarily equal $cHCO_3^-$. Thus the equilibrium equation for eq. 2 can be written:

$$K' = \frac{cH^+ . cHCO_3}{cH_2CO_3} = \frac{(cH^+)^2}{cH_2CO_3} \qquad \text{eq. 4}$$

The Henry's law constant defined by eq. 1 is:

$$K_H = \frac{cH_2CO_3}{pCO_2} \qquad \text{eq. 5}$$

which defines cH_2CO_3 as $K_H . pCO_2$, which can now be substituted in eq. 4.

$$K' = \frac{(cH^+)^2}{K_H . pCO_2} \qquad \text{eq. 6}$$

Rearranging gives:

$$cH^+ = (K_H K' pCO_2)^{\frac{1}{2}} \qquad \text{eq. 7}$$

Substituting the appropriate values of the equilibrium constants (Table 1) and using a CO_2 partial pressure (pCO_2) of 360 ppm, i.e. 3.6×10^{-4} atm, will yield a hydrogen ion (H^+) concentration of 2.4×10^{-6} mol l^{-1} or a pH of 5.6.

Sulphur dioxide (SO_2) is at much lower concentrations in the atmosphere, but it has a greater solubility and dissociation constant. We can set equations analogous to those for CO_2.

$$cSO_{2(g)} + cH_2O \rightleftharpoons cH_2SO_{3(aq)} \qquad \text{eq. 8}$$

$$cH_2SO_3(aq) \rightleftharpoons cH^+_{(aq)} + cHSO^-_{3(aq)} \qquad \text{eq. 9}$$

and once again rearranging gives:

**Box 2.12
Cont.**

$$cH^+ = (K_H K' pSO_2)^{\frac{1}{2}} \qquad\qquad \text{eq. 10}$$

If a small amount of SO_2 is present in the air at a concentration of 5×10^{-9} atm (not unreasonable over continental land masses), we can calculate a pH value of 4.85. So even low concentrations of SO_2 have a profound effect on droplet pH.

Table 1 Henry's law constants and first dissociation constants for atmospheric gases that undergo hydrolysis (25°C)

Gas	K_H (mol l^{-1} atm^{-1})	K' (mol l^{-1})
Sulphur dioxide	2.0	2.0×10^{-2}
Carbon dioxide	0.04	4.0×10^{-7}

2.9 Removal processes

So far, we have examined the sources of trace gases and pollutants in the atmosphere and the way in which they are chemically transformed. Now we need to look at the removal process to complete the source–reservoir–sink model of trace gases that we have adopted.

Our discussions have emphasised the importance of the OH radical as a key entity in initiating reactions in the atmosphere. Attack often occurs through hydrogen abstraction, and subsequent reactions with oxygen and nitrogen oxides (as illustrated in Box 2.11). This serves to remind us that the basic transformation that takes place in the atmosphere is oxidation. This is hardly unexpected in an atmosphere dominated by oxygen, /so we can argue that reactions within the atmosphere generally oxidise trace gases.

Oxidation of non-metallic elements yields acidic compounds and it is this that explains the great ease with which acidification occurs in the atmosphere. Carbon compounds can be oxidized to organic compounds, such as formic acid (HCOOH) or acetic acid (CH_3COOH) or, more completely, to carbonic acid (H_2CO_3, i.e. dissolved CO_2). Sulphur compounds can form H_2SO_4 and, in the case of some organosulphur compounds, methane sulphonic acid (CH_3SO_3H). Nitrogen compounds can ultimately be oxidised to HNO_3. The solubility of many of these compounds in water makes rainfall an effective mechanism for their removal from the atmosphere. The process is known as 'wet removal'.

It is important to note that, even in the absence of SO_2, atmospheric droplets will be acidic through the dissolution of CO_2 (Box 2.12). This has implications for the geochemistry of weathering (see Chapter 3). The SO_2, however, does make a substantial contribution to the acidity of droplets in the atmosphere. It can, so to speak, acidify rain (Box 2.12). However, let us consider the possibility of subsequent reactions that can cause even more severe acidification:

$$H_2O_{2(aq)} + HSO^-_{3(aq)} \rightarrow SO^{2-}_{4(aq)} + H^+_{(aq)} + H_2O_{(1)} \qquad\qquad \text{eq. 2.30}$$

$$O_{3(aq)} + HSO^-_{3(aq)} \rightarrow SO^{2-}_{4(aq)} + H^+_{(aq)} + O_{2(aq)} \qquad\qquad \text{eq. 2.31}$$

Hydrogen peroxide (H_2O_2) and O_3 are the natural strong oxidants present in rainwater. These oxidants can potentially oxidise nearly all the SO_2 in a parcel of air. Box 2.13 shows that under such conditions rainfall may well have pH values lower than 3. This illustrates the high acid concentrations possible in the atmosphere as trace pollutants are transferred from the gas phase to droplets. Liquid water in the atmosphere has a volume about a million times smaller than the gas phase; thus a substantial increase in concentration results from dissolution.

After the water falls to the Earth, further concentration enhancement can take place if it freezes as snow. When snow melts the dissolved ions are lost preferentially, as they tend to accumulate on the outside of ice grains which make up snowpacks. This means that at the earliest stages of melting it is the dissolved H_2SO_4 that comes out. Concentration factors of as much as 20-fold are possible. This has serious consequences for aquatic organisms, and especially their young, in the spring as the first snows thaw. It is not just acid rain, but acid rain amplified.

It is also possible for gaseous or particulate pollutants to be removed directly from the atmosphere to the surface of the Earth under a process known as dry deposition. This removal process may take place over land or the sea, but it is still termed 'dry deposition'. It is really a bit of a misnomer because the surfaces available for dry deposition are often most effective when they are wet.

Box 2.13

Removal of sulphur dioxide from an air parcel

A parcel of air over a rural area of an industrial continent would typically be expected to contain sulphur dioxide (SO_2) at a concentration of 5×10^{-9} atm. This means that a cubic metre of air contains 5×10^{-9} m^3 of SO_2. We can convert this to moles quite easily because a mole of gas occupies 0.0245 m^3 at 15°C and atmospheric pressure. Thus our cubic metre of air contains $5 \times 10^{-9}/0.0245 = 2.04 \times 10^{-7}$ mol of SO_2. In a rain-laden cloud we can expect one cubic metre to contain about 1 g of liquid water, i.e. 0.001 dm^3.

If the SO_2 were all removed into the droplet and oxidised to sulphuric acid (H_2SO_4), we would expect the 2.04×10^{-7} mol to dissolve in 0.001 dm^3 of liquid water, giving a liquid-phase concentration of 2.04×10^{-4} mol l^{-1}. The H_2SO_4 formed is a strong acid (see Box 2.5) so dissociates with the production of two protons under atmospheric conditions:

$$H_2SO_4 \rightarrow 2H^+ + SO^{2-}_4 \qquad\qquad \text{eq. 1}$$

Thus the proton concentration will be 4.08×10^{-4} mol l^{-1}, or the pH 3.4. Evaporation of water from the droplet and removal of further SO_2 as the droplet falls through air below the cloud can lead to even further declines in pH.

2.10   **Further reading**

Brimblecombe, P. (1995) *Air Composition and Chemistry*. Cambridge University Press, Cambridge.

Brimblecombe, P. (1987) *The Big Smoke*. Methuen, London.

Elsom, D.M. (1992) *Atmospheric Pollution*. Blackwell Scientific Publications, Oxford.

Wellburn, A. (1988) *Air Pollution and Acid Rain*. Longman, Harlow.

3 The terrestrial environment

3.1 The terrestrial environment, crust and material cycling

Terrestrial environments consist of solid (rocks, sediments and soils), liquid (rivers, lakes and groundwater) and biological (plants and animals) components. The chemistry of terrestrial environments is dominated by reactions between the Earth's crust and fluids in the hydrosphere and atmosphere.

The terrestrial environment is built on continental crust, a huge reservoir of igneous and metamorphic rock (mass of continental crust = 23.6×10^{24} g). This rock, often called crystalline basement, forms most of the continental crust. About 80% of this basement is covered by sedimentary rocks, which have an average thickness of 5 km. About 60% of these sedimentary rocks are mudrocks (clay minerals and quartz — SiO_2), with carbonates (limestones — calcium carbonate ($CaCO_3$) — and dolostones — $MgCa(CO_3)_2$) and sandstones (mainly quartz) accounting for most of the rest (Fig. 3.1).

Mud, silt and sandy sediments form mainly by weathering — the breakdown and alteration of solid rock. These sediments are usually transported by rivers to the oceans. In seawater, sediments sink to the seafloor, where physical processes and chemical reactions (collectively known as diagenesis) convert them to sedimentary rock. Eventually these rocks become land again, usually during mountain building.

The geological record shows that this material–transport mechanism has operated for at least 3.8 billion years. New sediments are derived either from older sedimentary rocks or from newly generated or ancient igneous and metamorphic rock. The average chemical composition of suspended sediment in rivers, sedimentary mudrock and the upper continental crust is quite similar (Table 3.1). This suggests that rivers represent an important pathway of material transport (Table 3.2) and that sedimentary mudrocks record crustal composition during material cycling.

This chapter focuses on components of the terrestrial environment which are chemically reactive. Natural water bodies — rivers, lakes and groundwater — fall into this category principally because water acts as a polar solvent (Box 3.1). Organisms, particularly plants and bacteria, may influence the types and rates of chemical reactions which occur in soils and water bodies. Humans can also influence some reactions, at times modifying them so profoundly that contamination or pollution occurs.

Some crustal solids are also reactive. Uranium (U) and potassium (K), common elements in granitic rocks, are inherently unstable because of their radioactivity (see Box 2.6). The radioactive decay of isotopes of uranium to form radon (Rn) gas can be a health hazard to humans living in areas with granitic bedrock (Box

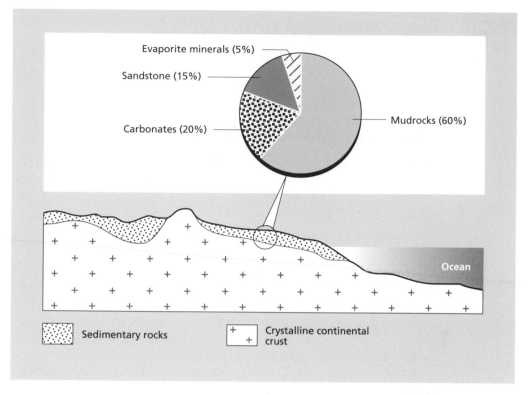

Fig. 3.1 Schematic cross-section of continental crust showing geometry and global average composition of sedimentary cover.

3.2). Some minerals are only stable under specific temperature and pressure conditions. For example, silicates which form deep in the crust, at high temperature and pressure, are unstable when exposed to Earth surface environments during weathering. The minerals adjust to the new set of conditions to regain stability. This adjustment may be rapid (minutes) for a soluble mineral such as halite (sodium chloride, NaCl) dissolving in water, or extremely slow (thousands or millions of years) in the weathering of silicates.

3.2 The structure of silicate minerals

Most of the Earth's crust is composed of silicate minerals (e.g. feldspars and quartz), which crystallised from magma or formed during metamorphism. Silicate minerals are compounds principally of silicon (Si) and oxygen (O), usually combined with other metals. The basic building block of silicates is the SiO_4 tetrahedron, in which silicon is situated at the centre of a tetrahedron of four oxygen ions (Fig. 3.2). This arrangement of ions is caused by the attraction — and strength of bonding — between positively charged and negatively charged ions (Box 3.3), and the relative size of the ions — which determines how closely neighbouring ions can approach one another (Section 3.2.1).

Box 3.1

Properties of water and hydrogen bonds

The water molecule H_2O is triangular in shape, with each hydrogen (H) bonded to the oxygen (O) as shown in Fig. 1. The shape results from the geometry of electron orbits involved in the bonding. Oxygen has a much higher electronegativity (see Box 3.4) than hydrogen and pulls the bonding electrons toward itself and away from the hydrogen atom. The oxygen thus carries a partial negative charge (usually expressed as $\delta-$), and each hydrogen a partial positive charge ($\delta+$), creating a dipole (i.e. electrical charges of equal magnitude and opposite sign a small distance apart). At any time a small proportion of water molecules dissociate completely to give H^+ and OH^- ions.

$$H_2O_{(l)} \rightleftharpoons H^+_{(aq)} + OH^-_{(aq)} \qquad \text{eq. 1}$$

for which the equilibrium constant is:

$$K_w = \frac{aH^+ . aOH^-}{aH_2O} = 10^{-14} \text{ mol}^2 \text{ l}^{-2} \qquad \text{eq. 2}$$

The activity of pure water is by convention unity, so eq. 2 simplifies to:

$$K_w = aH^+ . aOH^- = 10^{-14} \text{ mol}^2 \text{ l}^{-2} \qquad \text{eq. 3}$$

The polar nature of the water molecule allows the ions of individual water molecules to interact with their neighbours. The small hydrogen atom can approach and interact with the oxygen of a neighbouring molecule particularly effectively. The interaction between the hydrogen atom, with its partial positive charge, and oxygen atoms of neighbouring water molecules, with partial negative charges, is particularly strong — by the standards of intermolecular interactions — though weaker than within covalent bonding. This type of interaction is called hydrogen bonding.

The molecules in liquid water are less randomly arranged than in most liquids because of hydrogen bonds. The polarity of the bonds makes water an effective solvent for ions; the water molecules are attracted to the ion by electrostatic force to form a cluster around it. Moreover, ionic bonded compounds, with charge

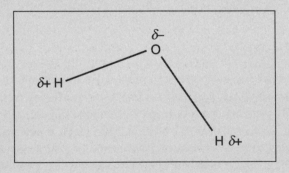

Fig. 1 The water molecule.

**Box 3.1
Cont.**

separation between component ions, are easily decomposed by the force of attraction of the water dipole. Hydrogen bonding gives water a relatively high viscosity and heat capacity in comparison with other solvents. Hydrogen bonds also allow water to exist as a liquid over a large temperature range. Since most biological transport systems are liquid, this latter property is fundamental to supporting life.

Box 3.2

Radon gas: a natural environmental hazard

Radon gas (Rn) is a radioactive decay product of uranium (U), an element present in crustal oxides (e.g. uraninite — UO_2), silicates (e.g. zircon — $ZrSiO_4$) and phosphates (e.g. apatite — $Ca_5(PO_4)_3$ (OH, F, Cl)). These minerals are common in granitic rocks, but are also to be found in other rocks, sediments and soils. Uranium decays to radium (Ra), which in turn decays to radon (Rn) (see Box 2.6). The isotope ^{222}Rn exists for just a few days before it also decays, but, if surface rocks and soils are permeable, this gas has time to migrate into caves, mines and houses. Here, radon or its radioactive decay products may be inhaled by humans. The initial decay products, isotopes of polonium, ^{218}Po and ^{216}Po, are non-gaseous and stick to particles in the air. When inhaled they lodge in the lungs' bronchi, where they decay — ultimately to stable isotopes of lead (Pb) — by ejecting α radiation particles (see Box 2.6) in all directions, including into the cells lining the bronchi. This radiation causes cell mutation and ultimately lung cancer. Having said this, radon is estimated to cause only about one in 20 cases of lung cancer in Britain, smoking being a much more serious cause.

Radon gas is invisible, odourless and tasteless. It is therefore difficult to detect and its danger is worsened by containment in buildings. Radon is responsible for about half the annual radiation dose to people in England, compared with < 1% from fallout, occupational exposures and discharges from nuclear power stations.

In England, about 100 000 homes are above the government-adopted 'action level' of 200 becquerels m^{-3}. Various relatively low-cost steps can be taken to minimise home radon levels, including better underfloor sealing and/or ventilation. Building homes in low-radon areas remains an obvious long-term strategy, but such simple solutions are not always applicable, because of either geographic or economic constraints. For example, bauxite processing in Jamaica produces large amounts of waste red mud. This material binds together strongly when dry, and is readily available as a cheap building material. Unfortunately, the red mud also contains higher levels of ^{238}U than most local soils. These cheap bricks are thus radioactive from the decay of ^{238}U and a potential source of radon. Only careful consideration of the health risks in comparison with the economic benefits can decide whether red mud will be used as a building material.

Table 3.1 Average chemical composition of upper continental crust, sedimentary mudrock and suspended load of rivers. Data from Taylor & McLennan (1985)

	Average upper continental crust (wt%)*	Average sedimentary mudrock† (wt%)	Average suspended load (rivers)‡ (wt%)
SiO_2	66.0	62.8	61.0
TiO_2	0.5	1.0	1.1
Al_2O_3	15.2	18.9	21.7
FeO	4.5	6.5	7.6
MgO	2.2	2.2	2.1
CaO	4.2	1.3	2.3
Na_2O	3.9	1.2	0.9
K_2O	3.4	3.7	2.7
Σ	99.9	99.9	99.4

* A silicate analysis is usually given in units of weight % of an oxide (grams of oxide per 100 g of sample). As most rocks consist mainly of oxygen-bearing minerals, this convention removes the need to report oxygen separately. The valency of each element governs the amount of oxygen combined with it. A good analysis should sum (Σ) to 100 wt%.
† This analysis represents terrigenous mudrock (i.e. does not include carbonate and evaporite components), a reasonable representation of material weathered from the upper continental crust.
‡ Average of Amazon, Congo, Ganges, Garronne and Mekong data.

Table 3.2 Agents of material transport to the oceans. After Garrels *et al.* (1975)

Agent	Percentage of total transport	Remarks
Rivers	89	Present dissolved load 17%, suspended load 72%. Present suspended load higher than geological past due to human activities (e.g. deforestation) and presence of soft glacial sediment cover
Glacier ice	7	Ground rock debris plus material up to boulder size. Mainly from Antarctica and Greenland. Distributed in seas by icebergs. Composition similar to average sediments
Groundwater	2	Dissolved materials similar to river composition. Estimate poorly constrained
Coastal erosion	1	Sediments eroded from cliffs, etc. by waves, tides, storms, etc. Composition similar to river suspended load
Volcanic	0.3 (?)	Dusts from explosive eruptions. Estimate poorly constrained
Wind-blown dust	0.2	Related to desert source areas and wind patterns, e.g. Sahara, major source for tropical Atlantic. Composition similar to average sedimentary rock. May have high (< 30%) organic matter content

Box 3.3

Ionic bonding, ions and ionic solids

In molecules like oxygen (O_2) and nitrogen (N_2), individual atoms bond by sharing electrons. However, many crystalline inorganic materials bond together by *donating and accepting* electrons, rather than sharing them. In fact, it can be argued that these structures have no bond at all, because the atoms entirely lose or gain electrons. This behaviour is usually referred to as ionic bonding. The classic example of an ionic solid is sodium chloride (NaCl).

$$Na. + :\overset{..}{\underset{..}{Cl}}\cdot \rightarrow Na^+ : \overset{..}{\underset{..}{Cl}}^- : \qquad \text{(dots represent electrons)} \qquad \text{eq. 1}$$

The theory behind this behaviour is that elements with electronic structures close to those of inert (noble) gases lose or gain electrons to achieve a stable (inert) structure. In eq. 1, sodium (Na; atomic no. = 11) loses one electron to attain the electronic structure of neon (Ne; atomic no. = 10), while chlorine (Cl; atomic no. = 17) gains one electron to attain the electronic structure of argon (Ar; atomic no. = 18). The compound NaCl is formed by the transfer of one electron from sodium to chlorine and bonded by the electrostatic attraction of the donated/received electron. The compound is electrically neutral.

Crystalline solids like NaCl are easily dissolved by polar solvents, such as water (see Box 3.1), which break down the ionic crystal into a solution of separate charged ions:

$$\overset{..}{Na^+ : \underset{..}{Cl}^-} : \overset{H_2O}{\rightleftharpoons} Na^+_{(aq)} + :\overset{..}{\underset{..}{Cl}}^-_{(aq)} \qquad \text{eq. 2}$$

(N.B. Most equations in this book will not show the electrons on individual ions and atoms.)

Positively charged atoms like Na^+ are known as cations, while negatively charged ions like Cl^- are called anions. Thus, metals whose atoms have one, two or three electrons more than an inert gas structure form monovalent (e.g. potassium — K^+), divalent (e.g. calcium — Ca^{2+}) or trivalent (e.g. aluminium — Al^{3+}) cations. Similarly, non-metals whose atoms have one, two or three electrons less than an inert gas structure form monovalent (e.g. bromine — Br^-), divalent (e.g. sulphur S^{2-}) and trivalent (e.g. nitrogen — N^{3-}) anions. In general, the addition or loss of more than three electrons is energetically unfavourable, and atoms requiring such transfers generally bond covalently (see Box 2.2).

The silicon ion Si^{4+} is an interesting exception. The high charge and small ionic radius make this cation polarising or electronegative (see Box 3.4), such that its bonds with the oxygen anion O^{2-} are distorted. This produces an appreciable degree of covalency in the Si–O bond.

When writing chemical equations, the sum of charges on one side of the equation must balance the sum of charges on the other side. On the left side of eq. 2, NaCl is an electrically neutral compound, whilst on the right side sodium and chloride each carry a single but opposite charge so that the charges cancel (neutralise) each other.

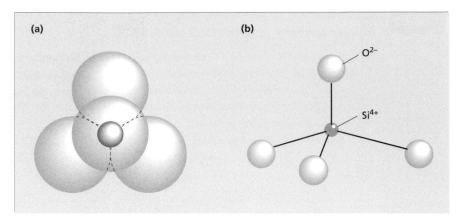

Fig. 3.2 Structure of SiO_4 tetrahedron. (a) Silicon and oxygen packing. The shaded silicon atom lies below the central oxygen atom, but above the three oxygens which lie in a single plane. (b) SiO_4 tetrahedron with bond length exaggerated.

3.2.1   Coordination of ions and the radius ratio rule

In crystals where bonding is largely ionic (Box 3.3), the densest possible packing of equal-sized anions (represented by spheres) is achieved by stacks of regular planar layers, as shown in Fig. 1. Spheres in a single layer have hexagonal symmetry, i.e. they are in symmetrical contact with six spheres. The layers are stacked such that each sphere fits into the depression between three other spheres in the layer below.

The gaps between neighbouring spheres have one of two possible three-dimensional geometries. The first geometry is delineated by the surfaces of four adjacent spheres. A prism constructed from the centre of each adjacent sphere (Fig. 3.3.) has the shape of a tetrahedron; consequently these gaps are called tetrahedral sites. The second type of gap is bounded by six adjacent spheres and a prism constructed from the centre of these spheres has the shape of a regular octahedron. These are called octahedral sites. In ionic crystals, cations occupy some of these tetrahedral and octahedral sites. The type of site a cation occupies is determined by the radius ratio of the cation and anion, i.e.:

$$\text{radius ratio} = r_{\text{cation}}/r_{\text{anion}} \qquad\qquad \text{eq. 3.1}$$

where r = ionic radius.

To fit exactly into an octahedral site delineated by six spheres of radius r, a cation must have a radius of $0.414\, r$ (called octahedral coordination). With this

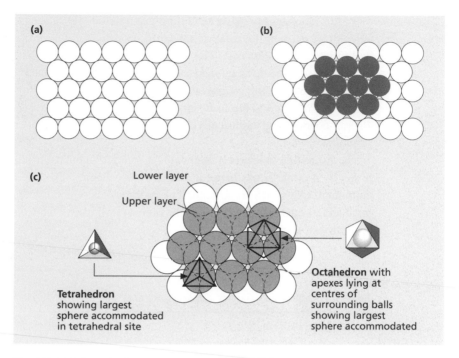

Fig. 3.3 (a) Spheres in planar layers showing hexagonal symmetry. (b) An upper layer of spheres (shaded) is stacked on the layer in (a), such that each upper sphere fits into the depression between three spheres in the lower layer. (c) Enlargement of (b), where heavy lines show coordination polyhedra, joining the centres of adjacent spheres, delineating two geometries, tetrahedra and octahedra. After McKie & McKie (1974) and Gill (1989).

radius ratio the cation touches all six of the surrounding anions and the short distance between ions means that the bond length is short and strong (optimum bond length). In real crystals, radius ratios are usually smaller or larger than 0.414. If smaller, the optimum bond length is exceeded, and the structure collapses into a new stable configuration where the cation maintains optimum bond length with fewer, more closely packed anions. If the radius ratio is larger than 0.414, octahedral coordination is maintained, but the larger cation prevents the anions from achieving their closest possible packing. The upper limit for octahedral coordination is the next critical radius ratio of 0.732, at which point the cation is large enough to simultaneously touch eight equidistant anion neighbours, reachieving optimum bond length.

In silicate minerals the layered stack of spheres is formed by oxygen anions (O^{2-}) and the radius ratio rule can be defined as:

$$\text{Radius ratio} = r_{cation}/r_{O^{2-}} \qquad\qquad \text{eq. 3.2}$$

Radius ratio values relative to O^{2-} are given in Table 3.3. The table shows that silicon (Si) exists in fourfold (tetrahedral) coordination with oxygen (O), i.e. it will fit into a tetrahedral site. This explains the existence of the SiO_4 tetrahedron. Octahedral sites, being larger than tetrahedral sites, accommodate cations of larger radius. However, some cations, e.g. strontium (Sr^{2+}) and caesium (Cs^+) (radius ratio > 0.732), are too big to fit into octahedral sites. They exist in eightfold or 12-fold coordination and usually require minerals to have an open, often cubic, structure.

The radius ratio rule is only applicable to ionic compounds. In silicate minerals, however, it is the bonds between oxygen and silicon and between oxygen and aluminium (Al) which are structurally important. These bonds are almost equally ionic and covalent in character and the radius ratio rule predicts the coordination of these ions adequately.

Table 3.3 Radius ratio values for cations relative to O^{2-}. From Raiswell *et al.* (1980)

Critical radius ratio	Predicted coordination	Ion	Radius ratio $r_k/r_{O^{2-}}$	Commonly observed coordination numbers
	3	C^{4+}	0.16	3
	3	B^{3+}	0.16	3, 4
0.225				
	4	Be^{2+}	0.25	4
	4	Si^{4+}	0.30	4
	4	Al^{3+}	0.36	4, 6
0.414				
	6	Fe^{3+}	0.46	6
	6	Mg^{2+}	0.47	6
	6	Li^+	0.49	6
	6	Fe^{2+}	0.53	6
	6	Na^+	0.69	6, 8
	6	Ca^{2+}	0.71	6, 8
0.732				
	8	Sr^{2+}	0.80	8
	8	K^+	0.95	8–12
	8	Ba^{2+}	0.96	8–12
1.000				
	12	Cs^+	1.19	12

Box 3.4

Electronegativity

Electronegativity is a measure of the tendency of an atom to attract an additional electron. It is used as an index of the covalent (see Box 2.2) or ionic (see Box 3.3) nature of bonding between two atoms. Atoms with identical electronegativity, or molecules such as nitrogen (N_2) consisting of two identical atoms, share their bonding electrons equally and so form pure covalent bonds (see Box 2.2). When component atoms in a compound are dissimilar, the bonds may become progressively polar. For example, in hydrogen chloride (HCl) the chlorine (Cl) atoms have a strong affinity for electrons, which are to a small extent attracted away from the hydrogen (H) toward chlorine. The bonding electrons are still shared, but not equally as in N_2.

H   : Cl   N   :   N   (: = bonding electrons)

(polar bond)   (covalent bond)

Hence the chlorine atom carries a slight negative charge and the hydrogen atom a slight positive charge. Extreme polarisation means that the bond becomes ionic in character (see Box 3.3). A bond is considered ionic if it has more than 50% ionic character.

Elements that donate electrons (e.g. magnesium, calcium, sodium and potassium) rather than attract them are called electropositive.

Table 1 Partial list of electronegativities and percentage ionic character of bonds with oxygen

Ion	Electro-negativity	% Ionic character	Ion	Electro-negativity	% Ionic character	Ion	Electro-negativity	% Ionic character
Cs^+	0.7	89	Zn^{2+}	1.7	63	P^{5+}	2.1	35
K^+	0.8	87	Sn^{2+}	1.8	73	Au^{2+}	2.4	62
Na^+	0.9	83	Pb^{2+}	1.8	72	Se^{2-}	2.4	—
Ba^{2+}	0.9	84	Fe^{2+}	1.8	69	C^{4+}	2.5	23
Li^+	1.0	82	Si^{4+}	1.8	48	S^{2-}	2.5	—
Ca^{2+}	1.0	79	Fe^{3+}	1.9	54	I^-	2.5	—
Mg^{2+}	1.2	71	Ag^+	1.9	71	N^{5+}	3.0	9
Be^{2+}	1.5	63	Cu^+	1.9	71	Cl^-	3.0	—
Al^{3+}	1.5	60	B^{3+}	2.0	43	O^{2-}	3.5	—
Mn^{2+}	1.5	72	Cu^{2+}	2.0	57	F^-	4.0	—

Measurements of % ionic character are not applicable for anions since their bonds with oxygen are predominantly covalent. A bond is considered ionic if it has more than 50% ionic character.

3.2.2   The construction of silicate minerals

The SiO_4 tetrahedron has a net 4⁻ charge, since silicon has a valency of 4⁺, and each oxygen is divalent (2⁻). This means that the silicon ion (Si^{4+}) can satisfy only half of the bonding capacity of its four oxygen neighbours. The remaining bonds are used in one of two ways as silicates crystallise from a magma.

1 Some magmas are rich in elements which are attracted to the electronegative tetrahedral oxygen (Box 3.4). The bonds between these elements (e.g. magnesium (Mg)) and oxygen have ionic character (Box 3.4) and result in simple crystal structures, e.g. olivine (the magnesium-rich form is called forsterite) (Section 3.2.3). The cohesion of forsterite relies on the Mg^{2+}–SiO_4^{4-} ionic bond. Bonding *within* the SiO_4 tetrahedron has a more covalent character. During weathering, water, a polar solvent (see Box 3.1), severs the weaker metal–SiO_4 tetrahedron ionic bond, rather than bonds within the tetrahedron itself. This releases metals and free SiO_4^{4-} as silicic acid (H_4SiO_4).

2 In some magmas, electropositive elements (opposite behaviour to electronegative elements) like magnesium are scarce. In these magmas each oxygen ion is likely to bond to two silicon ions, forming bonds of covalent character. The formation of extended networks of silicon–oxygen is called polymerisation, and is used to classify structural organisation in silicate minerals (Section 3.2.3).

The degree of structural complexity in a silicate mineral is a fundamental control on its reactivity and determines the way it behaves during Earth surface weathering.

3.2.3   Structural organisation in silicate minerals

Silicate minerals are classified by the degree to which silicon–oxygen bonded networks (polymers) form. Degree of polymerisation is measured by the number of non-bridging oxygens (i.e. those bonded to just one Si^{4+}).

Monomer silicates. These are built of isolated SiO_4 tetrahedra, bonded to metal cations as in olivine (Fig. 3.4b) and garnet. These minerals have four non-bridging oxygens and are also known as orthosilicates.

Chain silicates. If each SiO_4 tetrahedron shares two of its oxygens, chains of linked tetrahedra form (Fig. 3.4c). Chain silicates have two non-bridging oxygens and an overall Si : O ratio of 1 : 3, giving the general formula SiO_3. The pyroxene group of minerals provides the most important chain silicates — for example, enstatite ($MgSiO_3$). As in monomer silicates, bonding within chains is stronger than bonding between chains, which is between metal ions and non-bridging oxygens.

Fig. 3.4 (*Facing page*) (a) The schematic SiO_4 tetrahedron in Fig. 3.2a can be represented as a tetrahedron, each tip representing the position of oxygen anions. The sketches represent monomer structure in (b) and progressive polymerisation of adjacent tetrahedra to form (c) chains, (d) cross-linked double chains and (e) sheets. The cross-linked structures form hexagonal rings which can accommodate anions such as OH⁻. After Gill (1989).

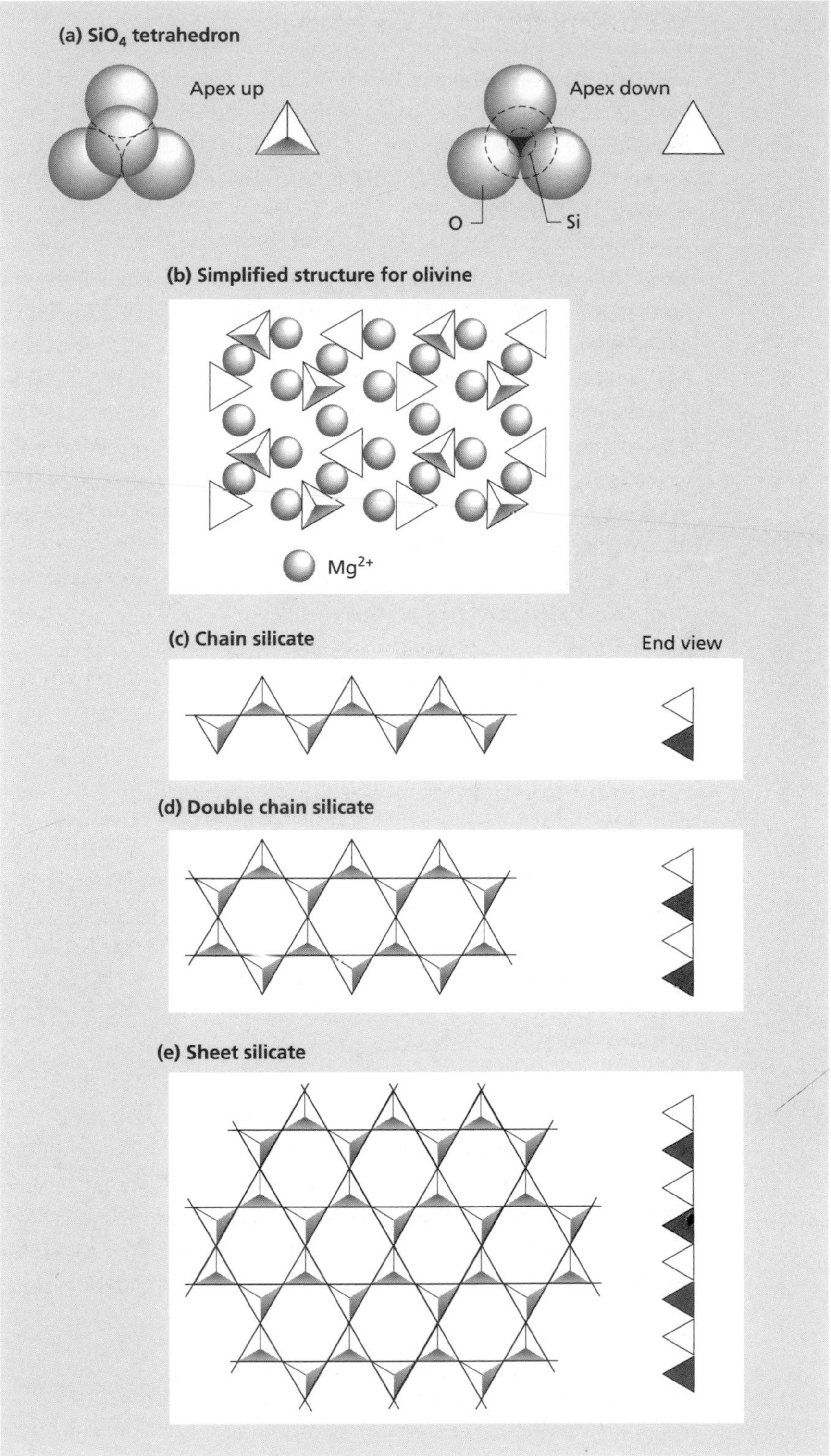

(a) SiO₄ tetrahedron

Apex up

Apex down

O — Si

(b) Simplified structure for olivine

Mg²⁺

(c) Chain silicate

End view

(d) Double chain silicate

(e) Sheet silicate

Double-chain silicates. In this structure the single chains are cross-linked, such that alternate tetrahedra share an oxygen with the neighbouring chain (Fig. 3.4d). Consequently, this structure has 1.5 non-bridging oxygens, since, for every four tetrahedra, two share two oxygens and the other two share three oxygens. The overall Si : O ratio is therefore 4 : 11, giving a general formula Si_4O_{11}. The amphibole group of minerals has double-chain structure — for example, tremolite $(Ca_2Mg_5Si_8O_{22}(OH)_2)$.

Sheet silicates. The next step in polymerisation is to cross-link chains into a continuous, semi-covalently bonded sheet, such that every tetrahedron shares three oxygens with neighbouring tetrahedra (Fig. 3.4e). This structure has one non-bridging oxygen and the overall Si : O ratio is 4 : 10, giving a general formula Si_4O_{10}. The hexagonal rings formed by the cross-linkage of chains are able to accommodate additional anions, usually hydroxide (OH^-). This structure is the basic framework for the mica group — for example, muscovite $(Mg_3(Si_4O_{10}(OH)_4)$ — and all of the clay minerals. These minerals are thus stacks of sheets, giving rise to their 'platy' appearance.

Framework silicates. In this class of silicates, every tetrahedral oxygen is shared between two tetrahedra, forming a three-dimensional semi covalent network. There are no non-bridging oxygens, the overall Si : O ratio being 1 : 2, as in the simplest mineral formula of the class, quartz (SiO_2). Substitution of aluminium (Al) into some of the tetrahedral sites (the ionic radius of aluminium is just small enough to fit) gives rise to a huge variety of aluminosilicate minerals, including the feldspar group, the most abundant mineral group in the crust. Substituting tetravalent silicon for trivalent aluminium causes a charge imbalance in the structure, which is neutralised by the incorporation of other divalent or monovalent cations. For example, in the feldspar, orthoclase ($KAlSi_3O_8$), one in four tetrahedral sites is occupied by aluminium in place of silicon. The charge is balanced by the incorporation of one K^+ for each tetrahedral aluminium.

3.3   Weathering processes

The surface of the continental crust is exposed to the atmosphere, making it vulnerable to physical and chemical processes. Physical weathering is a mechanical process which fragments rock into smaller particles without substantial change in chemical composition. When the confining pressure of the crust is removed by uplift and erosion, internal stresses within the underlying rocks are removed, allowing expansion cracks to open. These cracks may then be prised apart by thermal expansion (caused by diurnal fluctuations in temperature), by the expansion of water upon freezing and by the action of plant roots. Other physical processes, e.g. glacial activity, landslides and sandblasting, further weaken and break up solid rock. These processes are important since they vastly increase the surface area of rock material exposed to the agents of chemical weathering, i.e. air and water.

Chemical weathering is caused by water — particularly acidic water — and gases, e.g. oxygen, which attack minerals. Some ions and compounds of the original mineral are removed in solution, percolating through the mineral residue to feed groundwater and rivers. Fine-grained solids may be washed from the weathering site, leaving a chemically modified residue which forms the basis of soils. We can view weathering processes — physical and chemical weathering usually occur together — as the adjustment of rocks and minerals formed at high temperatures and pressures to Earth surface conditions of low temperature and pressure. The mineralogical changes occur to regain stability in a new environment.

3.4　Mechanisms of chemical weathering

Different mechanisms of chemical weathering are recognised and various combinations of these occur together during the breakdown of most rocks and minerals.

3.4.1　Dissolution

The simplest weathering reaction is the dissolution of soluble minerals. The water molecule (see Box 3.1) is effective in severing ionic bonds (see Box 3.3), such as those that hold sodium (Na^+) and chlorine (Cl^-) ions together in halite (rock salt). We can express the dissolution of halite in a simple way, i.e.

$$NaCl_{(s)} \underset{}{\overset{H_2O}{\rightleftharpoons}} Na^+_{(aq)} + Cl^-_{(aq)} \qquad \text{eq. 3.3}$$
(halite)

This reaction shows the dissociation (breaking of an entity into parts) of halite into free ions, forming an electrolyte solution. This reaction does not contain hydrogen ions (H^+), showing that the process is independent of pH.

3.4.2　Oxidation

Free oxygen is important in the decomposition of reduced materials (Box 3.5). For example, the oxidation of reduced iron (Fe^{2+}) and sulphur (S) in the common sulphide, pyrite (FeS_2), results in the formation of sulphuric acid (H_2SO_4), a strong acid.

$$2FeS_{2(s)} + 7\tfrac{1}{2}O_{2(g)} + 7H_2O_{(l)} \rightarrow 2Fe(OH)_{3(s)} + 4H_2SO_{4(aq)} \qquad \text{eq. 3.4}$$

Sulphides are common in sedimentary mudrocks, mineral veins and coal deposits. When mineral and coal deposits are mined, sulphide is left behind in the waste rock, which is piled in heaps. These spoil heaps have large surface areas exposed to the atmosphere where the oxidation of sulphides is extensive and rapid. In addition, abandoned mine workings are rapidly flooded by groundwater. The

Box 3.5

Redox reactions

Oxidation and reduction (redox) reactions are driven by electron transfers. Thus the oxidation of iron by oxygen:

$$4Fe_{(metal)} + 3O_{2(g)} \rightarrow 2Fe_2O_{3(s)} \qquad \text{eq. 1}$$

can be considered to consist of two half-reactions:

$$4Fe - 12e^- \rightarrow 4Fe^{3+} \qquad \text{eq. 2}$$

$$3O_2 + 12e^- \rightarrow 6O^{2-} \qquad \text{eq. 3}$$

where e^- represents one electron.

Oxidation involves the loss of electrons and *reduction involves the gain of electrons*. In eqs 1–3, oxygen — the oxidising agent (often called an electron acceptor) — is reduced because it gains electrons.

Equation 1 shows that elements like oxygen can exist under set conditions of temperature and pressure in more than one state, i.e. as oxygen gas and an oxide. The *oxidation state* of an element in a compound is assigned using the following rules.

1 The oxidation number of all elements is 0.

2 The oxidation number of a monatomic ion is equal to the charge on that ion, e.g. $Na^+ = Na(+1)$, $Al^{3+} = Al(+3)$, $Cl^- = Cl(-1)$.

3 Oxygen has an oxidation number of -2 in all compounds except O_2, peroxides and superoxides.

4 Hydrogen has an oxidation number of $+1$ in all compounds.

5 The sum of the oxidation numbers of the elements in a compound or ion equals the charge on that species.

6 The oxidation number of the elements in a covalent compound can be deduced by considering the shared electrons to belong exclusively to the more electronegative atom (Box 3.4). Where both atoms have the same electronegativity the electrons are considered to be equally shared. Thus the oxidation numbers of carbon and chloride in CCl_4 are $+4$ and -1 respectively and the oxidation number of chlorine in Cl_2 is 0.

Oxidation states are important when predicting the behaviour of elements or compounds. For example, chromium is quite insoluble and non-toxic as chromium (III), while as chromium (VI) it forms the soluble complex anion CrO_4^{2-}, which is toxic. As with most simple rules, those for oxidation state assignment apply to most but not all compounds.

Since redox half-reactions involve electron transfer, they can be measured electrochemically as electrode potentials, which are a measure of energy transfer (see Box 3.7). The reaction:

$$\tfrac{1}{2}H_2 - e^- \rightarrow H^+ \qquad \text{eq. 4}$$

is assigned an electrode potential (E°) of zero (at standard temperature and pressure) by international agreement. All other electrode potentials are measured

**Box 3.5
Cont.**

relative to this value (see Appendix 1, pp. 199–200). A positive $E°$ shows that the reaction proceeds spontaneously (e.g. the reduction of fluorine gas (oxidation state 0) to fluoride (F^-, oxidation state –1). A negative $E°$ shows that the reaction is spontaneous in the reverse direction (e.g. the oxidation of Li to Li^+).

To calculate the overall $E°$ for a reaction the relevant half-reactions are combined (regardless of the stoichiometry of the reactions). For example, the reaction of Sn^{2+} solution with Fe^{3+} solution involves two half-reactions:

$$Fe^{3+} + e^- \rightarrow Fe^{2+} \quad E° = 0.77 \text{ V} \qquad \text{eq. 5}$$

$$Sn^{4+} + 2e^- \rightarrow Sn^{2+} \quad E° = 0.15 \text{ V} \qquad \text{eq. 6}$$

These combine to give a positive $E°$, which shows that the forward reaction (eq. 7) is favoured.

$$2Fe^{3+} + Sn^{2+} \rightarrow 2Fe^{2+} + Sn^{4+} \qquad \text{eq. 7}$$

$E°$ for this reaction = 0.77 – 0.15 = 0.62 V.

The ability of any natural environment to bring about oxidation or reduction processes is measured by a quantity called its *redox potential* or Eh (see Box 3.15).

formation of H_2SO_4 makes drainage from abandoned mines strongly acidic (pH as low as 1 or 2). This acidity can increase aluminium solubility and cause toxicity in aquatic ecosystems (see Section 3.7.3). Microorganisms are closely involved in sulphide oxidation, which can be modelled by a series of reactions:

$$2FeS_{2(s)} + 2H_2O_{(l)} + 7O_{2(g)} \rightarrow 4H^+_{(aq)} + 4SO^{2-}_{4(aq)} + 2Fe^{2+}_{(aq)} \text{ (pyrite oxidation)} \qquad \text{eq. 3.5}$$

followed by the oxidation of ferrous iron (Fe(II)) to ferric iron (Fe(III)).

$$4Fe^{2+}_{(aq)} + O_{2(g)} + 10H_2O_{(l)} \rightarrow 4Fe(OH)_{3(s)} + 8H^+_{(aq)} \qquad \text{eq. 3.6}$$
$$\text{(Fe(II))} \hspace{4cm} \text{(Fe(III))}$$

Oxidation happens very slowly at the low pH values found in acid mine waters. However, below pH 3.5 iron oxidation is catalysed by the iron bacterium *Thiobaccillus thiooxidans* (Box 3.6). At pH 3.5–4.5 oxidation is catalysed by *Metallogenium*. Ferric iron may react further with pyrite.

$$FeS_{2(s)} + 14Fe^{3+}_{(aq)} + 8H_2O_{(l)} \rightarrow 15Fe^{2+}_{(aq)} + 2SO^{2-}_{4(aq)} + 16H^+_{(aq)} \qquad \text{eq. 3.7}$$

At pH values much above 3 the iron(III) precipitates as the common iron(III) oxide, goethite (FeOOH).

$$Fe^{3+}_{(aq)} + 2H_2O_{(l)} \rightarrow FeOOH_{(s)} + 3H^+_{(aq)} \qquad \text{eq. 3.8}$$

The precipitated goethite coats streambeds and brickwork as a distinctive yellow-orange crust.

Bacteria use iron compounds to obtain energy for their metabolic needs (e.g. oxidation of ferrous to ferric iron). Since these bacteria derive energy from the

Box 3.6

Reaction kinetics, activation energy and catalysts

Some reduced compounds appear to be stable at Earth surface temperatures despite the presence of atmospheric oxygen. Graphite, for example, is a reduced form of carbon which we might expect to react with oxygen, i.e.

$$C_{graphite} + O_{2(g)} \rightarrow CO_{2(g)} \qquad \text{eq. 1}$$

Although the reaction of oxygen with graphite is energetically favoured, graphite exists because the reaction is *kinetically* very slow. Many natural materials are out of equilibrium with their ambient environment and are reacting imperceptibly slowly. These materials are metastable. Metastability can be demonstrated using a graph of energy in a chemical system in which substances A and B react to give C and D (Fig. 1). In order for reaction to take place, A and B must come into close association and this usually requires an input of energy (activation energy). Under cold (low-energy) conditions a small number of A and B will occasionally have the energy to overcome the activation energy, but this will be rare and the reaction will proceed slowly. If the energy of the reactants is increased (for example, by heating), then the reaction will be able to proceed more quickly because more A and B will have the required activation energy.

An alternative way to increase the rate of a reaction is to lower the activation energy. This can be done by a catalyst, i.e. a substance that alters the rate of a reaction without itself changing. In our hypothetical reaction, a catalyst will allow A and B to come together more readily. In the environment, bacteria and their enzyme systems often catalyse reactions that would not otherwise proceed spontaneously because of kinetic inhibition. Water can also fulfil a catalytic role by allowing closer interactions of ions than is possible under dry conditions.

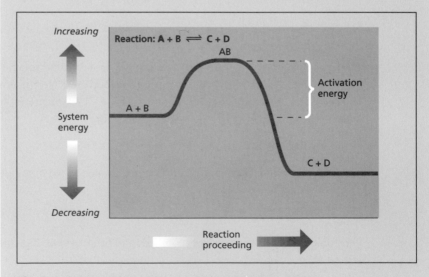

Fig. 1 Schematic representation of energy in a chemical system.

oxidation of inorganic matter, they thrive where organic matter is absent, using carbon dioxide (CO_2) as a carbon (C) source. Iron oxidation, however, is not an efficient means of obtaining energy; approximately 220 g of Fe^{2+} must be oxidised to produce 1 g of cell carbon. As a result, large deposits of iron(III) oxide form in areas where iron-oxidising bacteria survive.

In a mid-1980s survey, 10% of streams fed by groundwater springs in the northern Appalachians (USA) were found to be acidic due to acid mine drainage. Acidification of surface waters is likely to become worse in countries like the UK which are decommissioning large numbers of coalmines, often in restricted geographic areas, as a result of government energy policy.

Reduced iron-bearing silicate minerals, e.g. some olivines, pyroxenes and amphiboles, may also undergo oxidation, as depicted for the iron-rich olivine, fayalite.

$$Fe_2SiO_{4(s)} + \tfrac{1}{2}O_{2(g)} + 5H_2O_{(l)} \rightarrow 2Fe(OH)_{3(s)} + H_4SiO_{4(aq)} \qquad \text{eq. 3.9}$$

(fayalite)

(Fe(II))   (Fe(III))

The products are silicic acid (H_4SiO_4; see below) and colloidal hydrated iron oxide ($Fe(OH)_3$), a weak alkali, which dehydrates to yield a variety of iron oxides, e.g. Fe_2O_3 (haematite — dull red colour), FeOOH (goethite and lepidocrocite — yellow or orange-brown colour). The common occurrence of these iron oxides reflects their insolubility under oxidising Earth surface conditions.

The presence of water speeds up oxidation reactions, as shown by the everyday experience of metallic iron oxidation (rusting). Water acts as a catalyst, the oxidation potential being dependent on the partial pressure of gaseous oxygen and the acidity of the solution. At pH 7, water exposed to air has an Eh of 810 mV (see Box 3.5), an oxidation potential well above that necessary to oxidise ferrous iron.

Oxidation of organic matter. The oxidation of reduced organic matter in soils is catalysed by microorganisms. Bacterially mediated oxidation of dead organic matter to CO_2 is important because it generates acidity. In biologically active (biotic) soils, CO_2 may be concentrated by 10–100 times the amount expected from equilibrium with atmospheric CO_2, yielding carbonic acid (H_2CO_3) and H^+ via dissociation; to simplify the equations organic matter is represented by the generalised formula for carbohydrate, CH_2O.

$$CH_2O_{(s)} + O_{2(g)} \rightarrow CO_{2(g)} + H_2O_{(l)} \qquad \text{eq. 3.10}$$

$$CO_{2(g)} + H_2O_{(l)} \rightleftharpoons H_2CO_{3(aq)} \qquad \text{eq. 3.11}$$

$$H_2CO_{3(aq)} \rightleftharpoons H^+_{(aq)} + HCO^-_{3(aq)} \qquad \text{eq. 3.12}$$

These reactions may lower soil water pH from 5.6 (its equilibrium value with atmospheric CO_2 (see Box 2.12) to 4–5. This is a simplification since soil organic matter (humus) is not often completely degraded to CO_2. The partial breakdown products, however, possess carboxyl (COOH) or phenolic ($^{R}\!\bigcirc^{OH}$) groups, which dissociate to yield H^+ ions:

$$RCOOH_{(s)} \rightarrow RCOO^-_{(aq)} + H^+_{(aq)} \qquad \text{eq. 3.13}$$

$$^{R}\!\bigcirc^{OH}_{(s)} \rightarrow ^{R}\!\bigcirc^{O^-} + H^+_{(aq)} \qquad \text{eq. 3.14}$$

where R denotes a large organic entity. The acidity generated by organic matter decomposition is used to break down most silicate minerals by the process of acid hydrolysis.

3.4.3   Acid hydrolysis

Continental water contains dissolved species which render it acidic. The acidity comes from a variety of sources: from the dissociation of atmospheric CO_2 in rainwater — and particularly from dissociation of soil-zone CO_2 (see Section 3.4.2) — to form H_2CO_3, and natural and anthropogenic sulphur dioxide (SO_2) to form H_2SO_3 and H_2SO_4 (see Boxes 2.12 and 2.13). Reaction between a mineral and acidic weathering agents is usually called acid hydrolysis. The weathering of $CaCO_3$ demonstrates the chemical principle involved.

$$CaCO_{3(s)} + H_2CO_{3(aq)} \rightleftharpoons Ca^{2+}_{(aq)} + 2HCO^-_{3(aq)} \qquad \text{eq. 3.15}$$

The ionic Ca–CO_3 bond in the calcite crystal is severed and the released CO_3^{2-} anion attracts enough H^+ away from the H_2CO_3 to form the stable bicarbonate ion HCO_3^-. Note that the second HCO_3^- formed in eq. 3.15 is left over when H^+ is removed from H_2CO_3. Bicarbonate is a very weak acid, since it dissociates very slightly into H^+ and CO_3^{2-}, but it is not quite dissociated enough to react with carbonate. Overall, the reaction neutralises the acid contained in water. The reaction is dependent on the amount of CO_2 available: adding CO_2 causes the formation of more H_2CO_3 (eq. 3.15), which dissolves more $CaCO_3$ (forward reaction); conversely, lowering the amount of CO_2 encourages the reverse reaction and precipitation of $CaCO_3$. Stalactites and stalagmites forming in caves are an example of $CaCO_3$ precipitation induced by the degassing of CO_2 from groundwater. This response to varying CO_2 is a clear example of Le Chatelier's Principle (see Box 2.4).

Acid hydrolysis of a simple silicate, e.g. the magnesium-rich olivine, forsterite, is summarised by:

$$Mg_2SiO_{4(s)} + 4H_2CO_{3(aq)} \rightarrow 2Mg^{2+}_{(aq)} + 4HCO^-_{3(aq)} + H_4SiO_{4(aq)} \qquad \text{eq. 3.16}$$

Note that the dissociation of H_2CO_3 forms the ionised HCO_3^-, which is a slightly stronger acid than the neutral molecule (H_4SiO_4) donated by the destruction of the silicate.

The combined effects of dissolving CO_2 into soilwater (eq. 3.11), the subsequent dissociation of H_2CO_3 (eq. 3.12) and the production of HCO_3^- by acid hydrolysis weathering reactions (eqs 3.15 and 3.16) mean that surface waters have near-neutral pH, with HCO_3^- as the major anion.

3.4.4   Weathering of complex silicate minerals

So far we have considered the weathering of monomer silicates (e.g. olivine), which dissolve completely (congruent solution). This has simplified the chemical reactions. The presence of altered mineral residues during weathering, however, suggests that incomplete dissolution is more usual. Upper crustal rocks have an average composition similar to the rock granodiorite (Table 3.4). This rock is composed of the framework silicates, plagioclase feldspar, potassium feldspar and quartz (Table 3.4), plagioclase feldspar being most abundant. Thus, a simplified weathering reaction for plagioclase feldspar might best represent average chemical weathering. We can illustrate this using the calcium (Ca)-rich plagioclase feldspar, anorthite.

$$CaAl_2Si_2O_{8(s)} + 2H_2CO_{3(aq)} + H_2O_{(1)} \rightarrow Ca^{2+}_{3(aq)} + 2HCO^-_{3(aq)}$$
$$+ Al_2Si_2O_5(OH)_{4(s)} \qquad \text{eq. 3.17}$$

The formula of the solid product $(Al_2Si_2O_5(OH)_4)$ is that for kaolinite, an important member of the serpentine–kaolin group of clay minerals (see Section 3.6.3). This reaction demonstrates incongruent dissolution of feldspar, i.e. dissolution with *in situ* reprecipitation of some compounds from the weathered mineral.

Schematic representation of the anorthite chemical weathering reaction (Fig. 3.5) shows the edge of the anorthite crystal in contact with an H_2CO_3 weathering solution. Natural crystal surfaces have areas of excess electrical charge caused by imperfections (rows of atoms slightly out of place) or damage (broken bonds). Areas of excess negative charge are preferentially attacked by soil acids, resulting in the formation of etch pits on the mineral surface (Fig. 3.6). Hydrogen ions, dissociated from H_2CO_3 hydrate the silicate surface. The ionic bonds between Ca^{2+} and SiO_4 tetrahedra are easily severed, releasing Ca^{2+} into solution. The result is a metal-deficient hydrated silicate and a calcium bicarbonate $(Ca^{2+} + 2HCO^-_3)$

Table 3.4 Percentage mineral composition of the upper continental crust. After *Geochimica Cosmochimica Acta*, **48**, Nesbit & Young, p. 1534. Copyright 1984, with kind permission from Elsevier Science Ltd, The Boulevard, Langford Lane, Kidlington OX5 1GB, UK

	Average upper continental crust	Average exposed continental crust surface
Plagioclase feldspar	39.9	34.9
Potassium feldspar	12.9	11.3
Quartz	23.2	20.3
Volcanic glass	—	12.5
Amphibole	2.1	1.8
Biotite mica	8.7	7.6
Muscovite mica	5.0	4.4
Chlorite	2.2	1.9
Pyroxene	1.4	1.2
Olivine	0.2	0.2
Oxides	1.6	1.4
Others	3.0	2.6

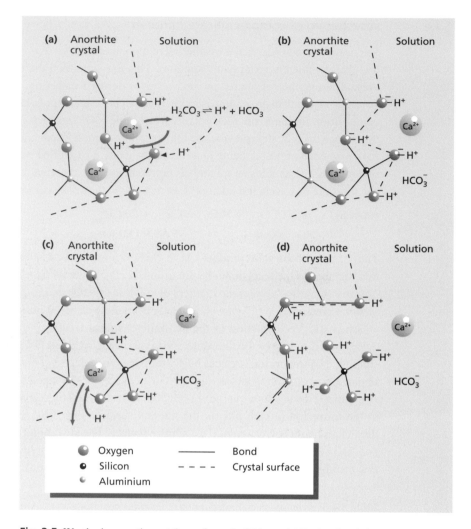

Fig. 3.5 Weathering reactions at the surface of a feldspar. (a) Broken bonds become protonated by H^+ dissociated from carbonic acid and ionic-bonded Ca^{2+} is released to solution. (b) Protonated lattice. (c) Further severing of ionic bonds causes complete protonation of the edge tetrahedron. (d) Edge tetrahedron is completely removed to solution as H_4SiO_4.

solution. Continued reaction may break the more covalent bonds within the tetrahedral framework. The tetrahedral framework is particularly weak where aluminium has substituted for silicon, since the aluminium–oxygen bond has more ionic character. The product released to solution is H_2SiO_4 (see Fig. 3.5). Equation 3.18 expresses quantitatively the reaction for the sodium (Na)-rich feldspar, albite.

$$2NaAlSi_3O_{8(s)} + 9H_2O_{(l)} + 2H_2CO_{3(aq)} \rightarrow Al_2Si_2O_5(OH)_{4(s)} + 2Na^+_{(aq)} + 2HCO^-_{3(aq)}$$
$$+ 4H_4SiO_{4(aq)} \qquad \text{eq. 3.18}$$

We might conclude that the dominant weathering mechanism of the upper crust is acid hydrolysis, resulting in a partially degraded and hydrated residue, with silicic acid and metal bicarbonate dissolved in water. We should, however, remember that much of the Earth's continental area is mantled by younger sedimentary rocks, including quite soluble ones, e.g. limestone. Limestones are common in young mountain belts, such as the European Alps and the Himalayas, where rates of physical weathering are high. It is therefore probable that average weathering reactions are really biased toward the weathering of the sediment cover, rather than average continental crust. Studies of Alpine weathering seem to confirm this. Many Alpine streamwaters are low in dissolved sodium and H_4SiO_4 but high in calcium and HCO_3^-. These results suggest that the dissolution of limestone — not feldspar — is the locally important weathering reaction.

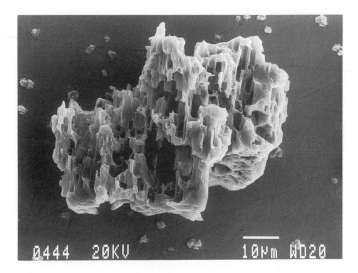

Fig. 3.6 Scanning electron micrograph showing square-shaped etch pits developed on dislocations in a feldspar from a southwestern England granite. Note that in places the pits are coalescing, causing complete dissolution of the feldspar. Scale bar = 10 μm. Photograph courtesy of ECC International, St Austell, UK.

3.5 Rate controls on weathering reactions

We have seen that the oxidation of soil organic matter causes acidity of natural waters, which promotes chemical weathering. This implies that the biosphere has an influence on the rate of weathering reactions. Other important factors include the relief of land areas, climate — principally rainfall and temperature — water composition, the type of parent material (bedrock) and the reaction kinetics of specific minerals. For clarity, these factors are discussed individually although in nature they operate together.

3.5.1 Temperature and water flow rate

Heat speeds up chemical reactions by supplying energy (Box 3.7). For most

Box 3.7

Chemical energy

The study of energy change is called thermodynamics. For example, the combustion of graphite carbon yields energy.

$$C_{graphite} + O_{2(g)} \rightarrow CO_{2(g)} \qquad \text{eq. 1}$$

The total energy released, or the energy change in going from reactants to products, is termed the change in Gibbs free energy (ΔG), which is measured in kilojoules per mole ($kJ \, mol^{-1}$). If energy is released, i.e. the products have lower free energy than the reactants, ΔG is considered negative. $\Delta G°$ for the burning of graphite under standard temperature (25°C) and pressure (1 atm), indicated by the superscript °, is −394.4 $kJ \, mol^{-1}$. Tables of $\Delta G°$ for various reactions are available and values of $\Delta G°$ for different reactions can be calculated by simple arithmetic combination of tabulated values. Any reaction with a negative ΔG value will in theory proceed spontaneously — the chemical equivalent of water flowing downhill — releasing energy. The reverse reaction requires an input of energy, i.e.

$$CO_{2(g)} \rightarrow C_{(graphite)} + O_{2(g)} \qquad \Delta G° = +394.4 \, kJ \, mol^{-1} \qquad \text{eq. 2}$$

Since an energetically favoured reaction proceeds from reactants to products, there is a relationship between ΔG and the equilibrium constant (K) for a reaction.

$$\Delta G = -RT \ln K \qquad \text{eq. 3}$$

where T is the absolute temperature (measured in kelvin (K)) and R is the universal gas constant (8.314 $J \, mol^{-1} \, K^{-1}$), relating pressure, volume and temperature for an ideal gas (see Box 2.3).

Converting eq. 3 to decimal logarithms gives:

$$\Delta G = -RT \, 2.303 \log_{10} K \qquad \text{eq. 4}$$

which at 25°C (298 K) yields:

$$\Delta G° = -5.707 \log_{10} K \qquad \text{eq. 5}$$

or:

$$\log_{10} K = \frac{-\Delta G°}{5.707} \qquad \text{eq. 6}$$

The total energy released in a chemical reaction has two components, enthalpy and entropy. Change in enthalpy (ΔH, measured in $J \, mol^{-1}$) is a direct measure of the energy emitted or absorbed by a reaction. Change in entropy (ΔS, measured in $J \, mol^{-1} \, K^{-1}$) is a measure of the degree of disorder. Most reactions proceed to increase disorder, e.g. by splitting a compound into constituent ions. Enthalpy and entropy are related:

$$\Delta G = \Delta H - T \Delta S \qquad \text{eq. 7}$$

Box 3.7
Cont.

In most reactions the enthalpy term dominates, but in some reactions the entropy term is important. For example, the dissolution of the soluble fertiliser potassium nitrate (KNO_3) occurs spontaneously. However, ΔH for the reaction:

$$KNO_{3(s)} \rightarrow K^+_{(aq)} + NO^-_{3(aq)} \qquad \text{eq. 8}$$

is +35 kJ and the solution absorbs heat (gets colder) as KNO_3 dissolves. Despite the positive enthalpy, the large increase in disorder (entropy) in moving from a crystalline solid to ions in a solution, gives an overall favourable energy balance or negative ΔG for the reaction.

Electrode potentials ($E°$; see Box 3.5) are a measure of energy transfer and so can be related to G:

$$G° = -nFE° \qquad \text{eq. 9}$$

where n is the number of electrons transferred and F is the universal Faraday constant (the quantity of electricity equivalent to one mole of electrons $= 6.02 \times 10^{23}$ e$^-$).

reactions a 10°C rise in temperature causes at least a doubling of reaction rate. This suggests that weathering in the tropics, where mean annual temperatures are around 20°C, will be about double the rate of weathering in temperate regions where mean annual temperatures are around 12°C.

The effect of temperature is linked to the availability of water. The dry air of hot, arid environments is an ineffective weathering agent. Vegetation and hence soil organic matter are sparse, and this reduces the concentration of organic acids. Moreover, close contact between rock particles and acid is prevented by the lack of water. Short-lived rainfall events may move surface salts into the soil, but the general dominance of evaporation over rainfall means that soluble salts tend to precipitate on the land surface, forming crusts of gypsum, carbonate and other evaporite minerals.

In humid, tropical climates, weathering is rapid, partly because the high temperatures speed up reactions, but mainly because the consistent supply of heavy rainfall allows rapid flushing and removal of all but the most insoluble compounds, e.g. oxides of aluminium and iron. Flushing constantly removes (leaches) soluble components and is particularly important in the undersaturated zone of soils.

3.5.2   Mineral reaction kinetics and solution saturation

The discussion above suggests that the weathering rate of minerals is proportional to the water flow rate, but this is only true if the waters are close to saturation (Box 3.8) with respect to the weathering mineral. If water flow is continuous and sufficiently high, a limit is reached beyond which further flushing is no longer a rate-controlling factor.

With insoluble minerals (solubility $< 10^{-4}$ mol l^{-1} (Box 3.8)), including all silicates and carbonates, ion detachment from mineral surfaces is very slow, such that

Box 3.8

Solubility product, mineral solubility and saturation index

The dynamic equilibrium between a mineral and its saturated solution (i.e. the point at which no more mineral will dissolve), for example:

$$CaCO_{3(calcite)} \rightleftharpoons Ca^{2+}_{(aq)} + CO^{2-}_{3(aq)} \qquad \text{eq. 1}$$

is quantified by the equilibrium constant (K), in this case:

$$K = \frac{aCa^{2+}.aCO_3^{2-}}{aCaCO_3} \qquad \text{eq. 2}$$

Since the $CaCO_3$ is a solid crystal of calcite, it is difficult to express its presence in terms of activity (see Box 4.4). This is overcome by recognising that reaction between a solid and its saturated solution is not affected by the amount of solid surface presented to the solution (as long as the mixture is well stirred). Thus the activity of the solid is effectively constant; it is assigned a value of 1 or unity (see eq. 3), and makes no contribution to the value of K in eq. 3.

The equilibrium constant for a reaction between a solid and its saturated solution is known as the solubility product and is usually given the notation K_{sp}. Solubility products have been calculated for many minerals, usually using pure water under standard conditions (1 atm pressure, 25°C temperature).

The solubility product for calcite (eq. 1) is thus:

$$K_{sp} = \frac{aCa^{2+}.aCO_3^{2-}}{1} = aCa^{2+}.aCO_3^{2-} = 4.5 \times 10^{-9} \text{ mol}^2 \text{ l}^{-2} \qquad \text{eq. 3}$$

The solubility product can be used to calculate the solubility (mol l^{-1}) of a mineral in pure water. The case for calcite is simple since each mole of $CaCO_3$ that dissolves produces one mole of Ca^{2+} and one mole of CO_3^{2-}. Thus:

$$\text{Calcite solubility} = cCa^{2+} = cCO_3^{2-} \qquad \text{eq. 4}$$

and therefore:

$$(\text{Calcite solubility})^2 = cCa^{2+}.cCO_3^{2-} = K_{sp} = 4.5 \times 10^{-9} \text{ mol}^2 \text{ l}^{-2} \qquad \text{eq. 5}$$

Thus:

$$\text{Calcite solubility} = \sqrt{4.5 \times 10^{-9}} = 6.7 \times 10^{-5} \text{ mol l}^{-1} \qquad \text{eq. 6}$$

The degree to which a mineral has dissolved in water can be calculated using the saturation index, i.e.:

$$\text{Degree of saturation } \Omega = \frac{\text{IAP}}{K_{sp}} \qquad \text{eq. 7}$$

IAP is the ion activity product, i.e. the numerical product of ion activity in the water. An Ω value of 1 indicates saturation, values > 1 indicate supersaturation and values < 1 indicate undersaturation.

Box 3.8
Cont.

Groundwater in the Cretaceous chalk aquifer of Norfolk, UK, has a calcium ion (Ca^{2+}) activity of 1×10^{-3} mol l^{-1} and a carbonate ion (CO_3^{2-}) activity of 3.5×10^{-6} mol l^{-1}. The saturation state of the water with respect to calcite is:

$$\Omega = \frac{aCa^{2+}.aCO_3^{2-}}{K_{sp(calcite)}} = \frac{1 \times 10^{-3}. \, 3.5 \times 10^{-6}}{4.5 \times 10^{-9}} = 0.78 \qquad \text{eq. 8}$$

i.e. the water is undersaturated with respect to calcite.

ions never build up in solution close to the crystal surface. The weathering rate of these minerals thus depends mainly on the rate of ion detachment from the crystal surface, rather than the efficiency of flushing (water flow rate). Only in very soluble minerals (solubility $> 2 \times 10^{-3}$ mol l^{-1}), e.g. evaporite minerals, can ions detach rapidly from the mineral surface to form a microenvironment close to the crystal surface which is saturated with respect to the dissolving mineral. The rate of dissolution is then controlled by the efficiency of dispersal of these ions and water-flushing effects are important.

3.5.3 Type of parent material (bedrock)

Rates of rock weathering are strongly dependent on the solubility and stability of the constituent minerals. In silicate minerals, the degree of polymerisation of the tetrahedral units controls relative stability. Thus, susceptibility to weathering follows a sequence which is roughly the reverse of the original crystallisation order or 'Bowen's reaction series' (Fig. 3.7). High-temperature monomer silicates (e.g. olivines) with ionic metal–oxygen bonds are easily weathered, whereas framework silicates, such as quartz, are resistant.

There is experimental evidence that dissolution rates of specific monomer silicates (e.g. Ca_2SiO_4, Mg_2SiO_4, etc.) are proportional to the rate of reaction between the divalent cation and water molecules during hydration (see Section 3.7.1). The rate of reaction between water molecules and alkaline earth ions is related to ionic size ($Ca(H_2O)_6^{2+} > Mg(H_2O)_6^{2+} > Be(H_2O)_6^{2+}$). This is mirrored by experimental dissolution rates, where $Ca_2SiO_4 > Mg_2SiO_4 > Be_2SiO_4$, and is controlled by the relative strength of the cation–oxygen bond.

3.5.4 Soils and biology

We saw in Section 3.4.2 that the presence of soil organic matter and its decomposition by microorganisms greatly increase the CO_2 concentration of soilwater, making it acidic. The presence of soil itself also affects weathering rates. Soils can only form where plants help stabilise the substrate, preventing erosion by surface water or wind. Soils are composed of organic matter and small mineral particles, which present a large surface area to acidic soilwaters. The organic (humus) content of the soil improves water retention, keeping mineral surfaces and soilwater in close contact.

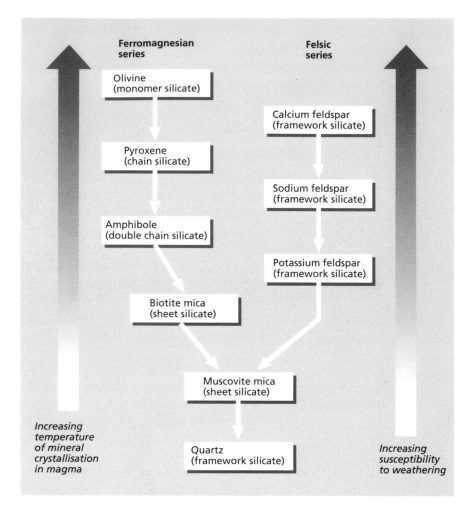

Fig. 3.7 Common silicate minerals ranked in Bowen's reaction series. Note that minerals formed under high temperatures with more ionic bonding are more susceptible to weathering. Ferromagnesium — minerals which contain essential iron and magnesium. Felsic — a rock containing feldspars and quartz.

There is debate about how effectively the presence of biotic soil influences weathering rates; some estimates suggest enhancements of 100–1000 times over abiotic weathering rates. The debate surrounding weathering rates is important, since the consumption of CO_2 by soil weathering reactions (see Section 3.4.3) lowers atmospheric partial pressure of CO_2 (pCO_2). Some researchers argue that, prior to the evolution of vascular land plants some 400 million years ago, weathering rates may have been much lower, giving rise to a higher atmospheric pCO_2 and enhanced greenhouse warming (see Section 5.3.4). Others, however, believe that thin soils, stabilised by primitive lichens and algae, covered the land surface billions of years before the evolution of vascular plants. These primitive biotic soils may have been quite effective in enhancing weathering rates, acting in a 'Gaian' way (see Section 1.3.3) by consuming atmospheric CO_2 and lowering global tempera-

Fig. 3.8 Spoil heap mainly of quartz, resulting from kaolinite extraction from highly weathered granite, Bodmin Moor, UK. Photograph courtesy of K. Clayton.

tures. This cooling effect may have helped improve the habitability of the early Earth for other organisms.

3.6 The solid products of weathering

The weathering of average upper-crustal granodiorite produces two types of solid product. Quartz is quite resistant to weathering and is concentrated in the residue (Figs 3.7 and 3.8). Although not strictly true, we will assume that quartz is chemically inert and takes no further part in chemical reactions. Feldspars, however, are weatherable and break down to form clay minerals.

3.6.1 Clay minerals

Clay minerals have a specific compositional character, being hydrous sheet silicates principally composed of oxygen, silicon and aluminium atoms. Clay minerals are small, $< 4\ \mu m$ in size (some would argue $< 2\ \mu m$), beyond the resolution of standard petrological microscopes. Sophisticated approaches, e.g. chemical, thermal, X-ray diffraction and scanning-electron microscopy, are needed to resolve clay mineral structure and composition.

3.6.2 Clay mineral composition

Clay minerals are sheet silicates (see Section 3.2.3) constructed of layers of atoms in tetrahedral and octahedral coordination, known as tetrahedral and octahedral sheets.

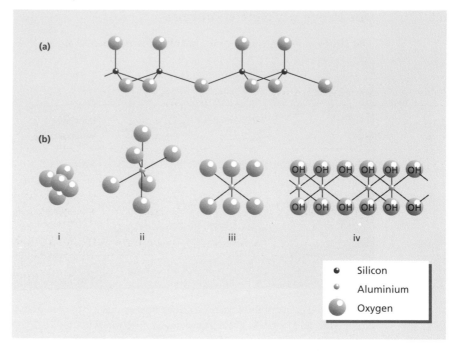

Fig. 3.9 (a) A sheet of SiO_4 tetrahedra linked via basal oxygens, with apical oxygens pointing upward. (b) Octahedra and the octahedral sheet: (i) the atoms packed together; (ii) the octahedron expanded; (iii) conventional representation of an octahedron; (iv) conventional representation of an octahedral sheet, showing aluminium equidistant between six hydroxyls — forming the mineral gibbsite.

The tetrahedral sheets are layers of SiO_4 tetrahedra which share three oxygens with neighbouring tetrahedra. These basal oxygens form a hexagonal pattern (see Section 3.2.3). The fourth tetrahedral (apical) oxygen of each tetrahedron is arranged perpendicular to the basal sheet (Fig. 3.9a). The sheet carries a net negative charge.

The octahedral sheet is composed of cations, usually aluminium, iron or magnesium, arranged equidistant from six oxygen (or OH) anions (Fig. 3.9b). Aluminium is the common cation and the ideal octahedral sheet has the composition of the aluminium hydroxide mineral, gibbsite ($Al(OH)_3$). Where octahedral sites are filled by trivalent aluminium, only two of every three sites are occupied to maintain electrical neutrality and the sheet is classified as dioctahedral (Table 3.5). Where divalent cations fill octahedral sites, all available sites are filled and the sheet is classified as trioctahedral (Table 3.5).

Combining these sheets gives the basic clay mineral structure. The combination allows the apical oxygen of the tetrahedral sheet and the OH groups lodged in the centre of the hexagonal holes of the basal tetrahedral sheet to be shared with the octahedral sheet (Fig. 3.10). The various clay mineral groups (Table 3.5) result from different styles of arrangement and mutual sharing of ions in the tetrahedral and octahedral sheets.

3.6.3   One to one clay mineral structure

The simplest arrangement of tetrahedral and octahedral sheets is a 1 : 1 layering,

Table 3.5 Simplified classification of clay minerals. After Martin *et al.* (1991)

Layer type	Group	Common minerals	Octahedral character	Interlayer material
1 : 1	Serpentine–Kaolin	Kaolinite	Dioctahedral	None
2 :1	Smectite	Montmorillonite	Dioctahedral	Hydrated exchangeable cations
	True (flexible) mica	Biotite	Trioctahedral	Non-hydrated monovalent cations
		Muscovite, illite	Dioctahedral	
	Chlorite	Chamosite	Trioctahedral	Hydroxide sheet

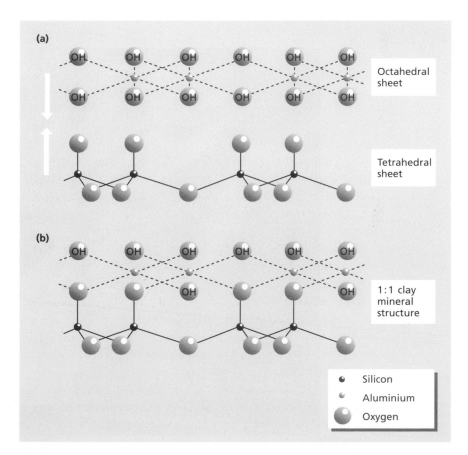

Fig. 3.10 Schematic diagram to show how the octahedral and tetrahedral sheets, seen as separate entities in (a), can be merged to form 1 : 1 clay mineral structure shown in (b).

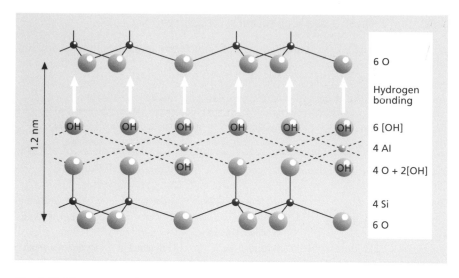

Fig. 3.11 The structure of a 1 : 1 clay mineral (kaolinite).

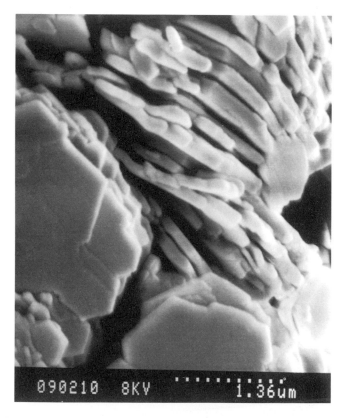

Fig. 3.12 Scanning electron microscope image of kaolinite showing plate-like crystals arranged in stacks. Scale bar = 1.36 μm. Photograph courtesy of S. Bennett.

shown in Fig. 3.10 and developed more fully in Fig. 3.11. These 1 : 1 minerals comprise the serpentine–kaolin group of clay minerals, of which the mineral kaolinite is probably the best known (Fig. 3.12). In kaolinite, the 1 : 1 packages are held together by hydrogen bonds (see Box 3.1), which bridge between OH groups of the upper layer of the octahedral sheet and the basal oxygens of the overlying tetrahedral sheet. The hydrogen bonds are strong enough to hold the 1 : 1 units together, preventing cations getting between layers (interlayer sites). Isomorphous substitution (Box 3.9) in kaolinite is negligible.

3.6.4   Two to one clay mineral structure

The other important structural arrangement is a 2 : 1 structure, comprising an octahedral layer, sandwiched between two tetrahedral sheets with apical oxygens pointing inward on each side of the octahedral sheet (Fig. 3.13). The mutual sharing of two layers of apical oxygens in the octahedral sheets implies a higher oxygen : OH ratio in the structure of the 2 : 1 vs. the 1 : 1 octahedral sheets. All of the other clay mineral groups share this structure, the most important being the mica group, which includes the common mica minerals and illite, the smectite group and the chlorite group.

Box 3.9

Isomorphous substitution

Isomorphism describes substances which have very similar structure. The carbonate mineral system is a good example, where some minerals differ only on the basis of the cation, for example, $CaCO_3$ (calcite), $MgCO_3$ (magnesite), $FeCO_3$ (siderite). The basic similarity of structure allows interchangeability of cations between end-member minerals. For example, most natural calcite has a measurable amount of both Mg^{2+} and Fe^{2+} substituted for some of the Ca^{2+}. The amount of isomorphous substitution is shown by the following notation, $(Ca_{0.85}Mg_{0.1}Fe_{0.05})CO_3$. In other words, 85% of the Ca^{2+} sites are occupied by Ca^{2+}, 10% of the Ca^{2+} sites are occupied by Mg^{2+}, and 5% of the Ca^{2+} sites are occupied by Fe^{2+}. The complex chemistry of freshwater and seawater means that natural minerals incorporate many trace elements and rarely conform to their ideal formulae.

The radius ratio rule (see Section 3.2.1) predicts that divalent Ca^{2+}, Mg^{2+} and Fe^{2+} will have sixfold coordination because of their similar ionic radii (0.106 nm Ca^{2+}, 0.078 nm Mg^{2+}, 0.082 nm Fe^{2+}). They are therefore interchangeable without upsetting either the physical packing or the electrical stability of an ionic compound.

In compounds where bonding has covalent character, isomorphous substitution is prevented. This is because the need for electron sharing in the bond modifies structures away from the simple packing geometries predicted by the radius ratio rule.

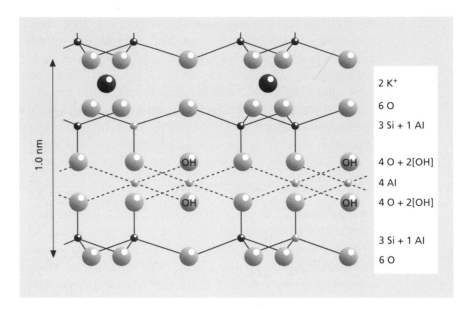

Fig. 3.13 The structure of muscovite mica.

Illite is a term used to describe clay-sized mica-type minerals and is not a specific mineral name; however, in general, illite composition is similar to muscovite mica (Fig. 3.13). In the muscovite structure, one of every four tetrahedral silicons is replaced by aluminium. The regular replacement of tetravalent silicon by trivalent aluminium means that the tetrahedral sheet in muscovite carries a strong net negative charge. Ideally, illite has dioctahedral structure, but some of the octahedral aluminium is substituted by Fe^{2+} and Mg^{2+} (Box 3.9), resulting in a net negative charge for the octahedral sheet. In total, the illite 2 : 1 unit has a strong net negative charge, known as the layer charge. This negative charge is neutralised by large cations, usually K^+, which sit between the 2 : 1 units and bond ionically with basal oxygens of the opposing tetrahedral sheets in sixfold coordination. The radius ratio rule predicts that K^+ should exist in eightfold or 12-fold coordination with oxygen (see Section 3.2.1), but this does not occur, due to slight distortion in the illite structure.

It is important to note that bonding between 2 : 1 illite units cannot be fulfilled by hydrogen bonds associated with OH groups (as in kaolinite), since each 2 : 1 unit presents only basal tetrahedral oxygens on its outer surfaces. Moreover, the ionic bonding between K^+ in the interlayer site and the tetrahedral oxygens is a relatively strong bond, making illitic clays stable minerals. This accounts for their abundance as weathering products, particularly in temperate and colder climates.

The smectite group of clay minerals are structurally similar to illites (Fig. 3.14). In octahedral sites, substitution of Al^{3+} by Mg^{2+} or Fe^{3+} is common and some substitution of Si^{4+} by Al^{3+} in the tetrahedral sites also occurs, resulting in a net negative layer charge. This charge is, however, only about one-third the strength of

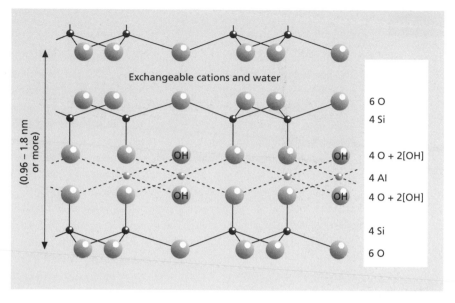

Exchangeable cations and water

(0.96 – 1.8 nm or more)

6 O
4 Si

4 O + 2[OH]

4 Al

4 O + 2[OH]

4 Si

6 O

Fig. 3.14 The structure of a 2 : 1 clay mineral (smectite).

the illite layer charge. Consequently, smectite is not able to bond interlayer cations effectively and the 2 : 1 units are not tightly bonded together. This allows water and other polar solvents to penetrate interlayer sites, causing the mineral to swell. Cations, principally H^+, Na^+, Ca^{2+} and Mg^{2+}, also enter the interlayer site with water and neutralise the negative charge. Bonding between the 2 : 1 units is effected by the hydrated cation interlayer, by a combination of hydrogen bonds and van der Waals' forces (Box 3.10). This weak bonding holds cations loosely in the interlayer sites, making them prone to replacement by other cations. As a consequence, smectites have high cation exchange capacity (see Section 3.6.6).

The similar structure of illite and smectite allows mixing or interstratification of 2 : 1 units to form mixed-layer clays. Most illites and smectites are interstratified to a small degree, but they are not classified as such until detectable by X-ray diffraction. As one might expect, illite–smectite mixed-layer clays have intermediate cation exchange capacity between the end-member compositions.

| 3.6.5 | Controls on clay mineral formation |

3.6.5 Controls on clay mineral formation

In an average upper-crustal granodiorite, it is mainly feldspars that weather to form clay minerals. Since feldspars are framework silicates, the production of sheet silicates (eqs 3.17 and 3.18) must involve an intermediate step. This step involves the release of silica, aluminium and other cations, followed by their recombination into a sheet silicate structure. Since this intermediate step involves ions in soil solutions, then ionic composition, soil water pH and the degree of leaching (water flow rate) will influence the type of clay mineral formed.

Box 3.10

Van der Waals' forces

Non-polar molecules have no permanent dipole and cannot form normal bonds. The non-polar noble gases, however, condense to liquid and ultimately form solids if cooled sufficiently. This suggests that some form of intermolecular force holds the molecules together in the liquid and solid state. The amount of energy required to melt solid xenon is 14.9 kJ mol^{-1}, demonstrating that cohesive forces operate between the molecules.

Weak, short-range forces of attraction, independent of normal bonding forces, are known as van der Waals' forces, after the nineteenth-century Dutch physicist. These forces arise because, at any particular moment, the electron cloud around a molecule is not perfectly symmetrical. In other words, there are more electrons (thus net negative charge) on one side of a molecule than on the other, generating an instantaneous electrical dipole. This dipole induces dipoles in neighbouring molecules, the negative pole of the original molecule attracting the positive pole of the neighbour. In this way, weak induced dipole–induced dipole attractions exist between molecules.

Induced dipoles continually arise and disappear as a result of electron movement, but the force between neighbouring dipoles is always attractive. Thus, although the average dipole on each molecule measured over time is zero, the resultant forces between molecules at any instant are not zero.

As the size of molecules increase, so do the number of constituent electrons. As a result, larger molecules have stronger induced dipole–induced dipole attractions. It must be stressed, however, that van der Waals' forces are much weaker than both covalent and ionic bonds.

Aluminium and iron precipitate as insoluble oxides or oxyhydroxides over the normal soil pH range. Other soil cations and H_4SiO_4 are quite soluble and thus prone to transport away from the weathering site. This difference in cation behaviour has been quantified as the chemical index of alteration (CIA); using molecular proportions:

$$CIA = \left(\frac{Al_2O_3}{Al_2O_3 + CaO^* + Na_2O + K_2O} \right) \times 100 \qquad \text{eq. 3.19}$$

where CaO* is CaO in silicate minerals (i.e. excludes Ca-bearing carbonates and phosphates). Thus, CIA values approaching 100 are typical of materials formed in heavily leached conditions where soluble calcium, sodium and potassium have been removed. Kaolinite clays attain such values (Table 3.6), whereas illites and smectites have CIA values around 75–85 (Table 3.6). In comparison, unleached feldspars have CIA values around 50.

The CIA predicts that kaolinite will form under heavily leached conditions, and this is confirmed by observations in tropical weathering regimes. On stable land surfaces where weathering and leaching have been prolonged, well-drained sites develop kaolinitic and, in extreme cases, gibbsitic clay mineralogies (Fig. 3.15). Such sites are mantled by iron-rich (laterite) and aluminous (bauxite) surface deposits (Fig. 3.16). These surface deposits can become thick enough to prevent further interaction between surface waters and bedrock, lowering the rate of subsequent bedrock weathering.

In contrast, smectite clays develop in poorly drained sites. On the basaltic island of Hawaii, soil clay mineral type changes in the sequence smectite–kaolinite–gibbsite as rainfall amount increases (Fig. 3.17). A similar, generalised zonation has been proposed for clay mineral distribution with depth in soils, again based on the degree of leaching (Fig. 3.18).

Intense leaching favours kaolinite formation since cations and H_4SiO_4 are removed, lowering the silicon : aluminium ratio and favouring the 1 : 1 structural arrangement. Less intense leaching favours a higher silicon : aluminium ratio, allowing the formation of various 2 : 1 clay minerals, depending on the supply of cations. For example, the weathering of basalt provides abundant magnesium for the formation of magnesium smectite. In the most intense tropical weathering environments, all of

Table 3.6 Chemical index of alteration (CIA) values for various crustal materials. Data from Nesbitt & Young (1982), Maynard *et al.* (1991) and Taylor & McLennan (1985)

Material	CIA
Clay minerals	
Kaolinite	100
Chlorite	100
Illite	75–85
Smectite	75–85
Other silicate minerals	
Plagioclase feldspar	50
Potassium feldspar	50
Muscovite mica	75
Sediments	
River Garonne (southern France) suspended load	75*
Barents Sea (silt)	65*
Mississippi delta average sediment	64*
Amazon delta muds	70–75
Amazon weathered residual soil clay	85–100
Rocks	
Average continental crust (granodiorite)	50
Average shales	70–75
Basalt	30–40
Granite	45–50

* Value calculated using total CaO rather than CaO* (see text).

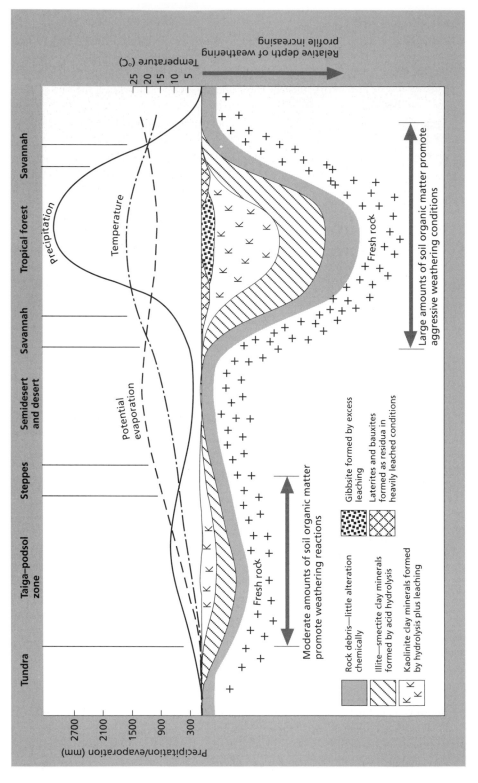

Fig. 3.15 Schematic relationship between weathering zones and latitude–vegetation–climatic zones. The influence of relief (mountainous areas), where soils are typically thin, is not included. After Strakhov (1967).

Fig. 3.16 Thick bauxitic soil (dark) overlying Tertiary limestone (white). The soil is piped into solution-enlarged hollows of the limestone surface. Cliff face approximately 6 m high. South Jamaica. Photograph courtesy of J. Andrews.

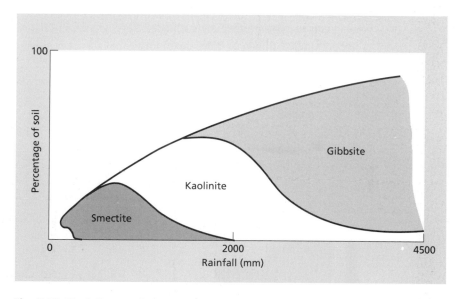

Fig. 3.17 The influence of climate on clay mineralogy in Hawaii. The relatively rapid water flow rates associated with high rainfall result in the preferential removal of cations and silica. After Sherman (1952).

the silica is removed, favouring the formation of gibbsite, which can be thought of as a 0 : 1 arrangement (i.e. only octahedral sheet present; Fig. 3.18).

It is not known how clay minerals form from solution. There is evidence that fulvic acids, from the decay of organic matter in soil, may react with aluminium to form a soluble aluminium–fulvic acid complex, with aluminium in sixfold coordination. This gibbsitic unit may then have SiO_4 tetrahedra adsorbed on to it to form the familiar clay mineral structures.

A minority of unweathered rock-forming silicates, for example the micas, are already sheet silicates. It is not difficult to envisage that alteration could transform these to clay minerals. Alteration is most likely in the interlayer areas, especially at damaged crystal edges. The clay mineral formed will depend on the composition of both the original mineral and the ions substituted during alteration. For example, the replacement of K^+ in muscovite by Mg^{2+} would lead to the formation of magnesium smectite.

3.6.6   Ion exchange in soils and the hydrosphere

Exchangeable ions are those that are held temporarily on materials by weak, electrostatic forces. If particles with one type of adsorbed ion are added to an electrolyte solution containing different ions, some of the particle-surface adsorbed ions are released into solution and replaced by those from the solution (Fig. 3.19).

We have seen that the interlayer sites of clay minerals, particularly smectites, hold ions weakly, giving these minerals a capacity for ion exchange. Clay mineral ion exchange can also be a surface phenomenon. Edge damage to minerals can break bonds to expose either uncoordinated oxygens (sites of net negative charge) or uncoordinated silicon or other metal ions (sites of net positive charge). These

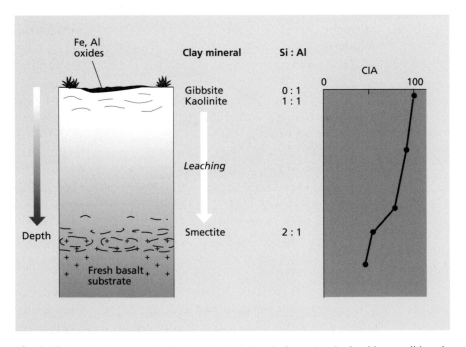

Fig. 3.18 Idealised vertical distribution of clay minerals formed under leaching conditions in soils developed on basalt. CIA values increase from 30–40 in fresh rock to near 100 in heavily leached surface soils.

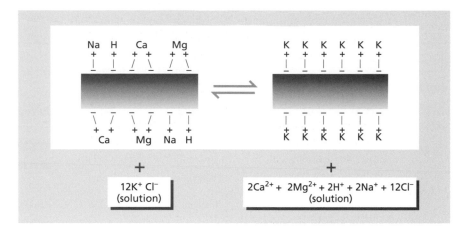

Fig. 3.19 Schematic diagram to show ion exchange equilibria on the surface of a clay particle. Potassium ions in solution are exchanged for other cations, causing the exchange equilibrium to move from left to right.

surface charges are balanced by electrostatic adsorption of cations and anions respectively. In addition, clays with damaged octahedral layers may expose OH groups. Under high pH conditions these OH groups may dissociate, forming a negative charge, which is neutralised by other cations, e.g.

$$\text{Clay–mineral–O}^-\text{–H}_{(s)} + \text{K}^+_{(aq)} \rightleftharpoons \text{clay–mineral–O}^-\text{–K}_{(s)} + \text{H}^+_{(aq)} \qquad \text{eq. 3.20}$$

Micrometre-sized clay minerals have a large surface area : volume ratio and constitute a significant sink for some anions and cations in soil environments. Despite this, in smectite clays, surface ion exchange is much less important than interlayer site exchange (Table 3.7).

It is often impossible to calculate the amount of ion exchange caused by clay minerals in soils, since other components, e.g. allophane and organic matter, have significant cation exchange capacity (Table 3.7). Allophane is a silicon–alluminium–water gel composed of 3–5-nm-diameter hollow spheres with walls composed of kaolinite-like material. The cation exchange sites are within the hollow spheres and exchange capacity is variable depending on the exchangeable cation, its concentration and soilwater pH. Ion exchange in soil organic matter is caused mainly by dissociation of carboxyl groups at pH above 5.

3.6.7 Use of clay minerals in cases of environmental contamination

The large cation exchange capacity of smectite has prompted research into its use as a catalyst, i.e. a substance that alters the rate of a chemical reaction without itself changing. Clay catalysts have potential applications as adsorbents to treat contaminated natural waters or soils.

The compound 2,3,7,8-tetrachlorodibenzo-*p*-dioxin is one of the most toxic priority pollutants on the US Environmental Protection Agency's list. Dioxin

Table 3.7 Representative cation exchange capacity, CEC (meq* $100g^{-1}$ dry weight), for various soil materials. After Birkeland (1974), by permission of Oxford University Press

Non-clay materials	CEC	Clay minerals	CEC	Cation exchange site
Quartz, feldspars	1–2			
Hydrous oxides of Al and Fe	4	Kaolinite	3–15	Edge effects
		Illite	10–40	Mainly edge effects, plus some interlayer
		Chlorite	10–40	
Allophane	20–100	Smectite	80–150	Mainly interlayer plus some edge effects
Organic matter	150–500			

* A milliequivalent (meq) is the charge carried by 1.008 milligrams (mg) of H^+, or the charge carried by any ion (measured in mg l^{-1}), divided by its relative atomic mass and multiplied by the numerical value of the charge. If, for example, divalent Ca^{2+} displaces H^+, it occupies the charged sites of $2H^+$ ions. Thus the amount of Ca^{2+} required to displace 1 meq of H^+ is 40 (atomic weight of Ca) divided by 2 (charge) = 20 mg, i.e. the weight of 1 meq of Ca^{2+}.

compounds act as nerve poisons and are extremely toxic. There is no lower limit at which dioxins are considered safe in natural environments. The destruction of dioxins by biological, chemical or thermal means is costly, not least because their low (but highly significant) concentrations are dispersed in large volumes of other (benign) material. Thus large volumes of material must be treated in dioxin destruction processes.

It is desirable to concentrate soluble contaminants like dioxin by adsorption on to a solid before destruction. The optimal solid adsorbent should be cheap, benign, recyclable and easy to handle and have a high affinity for — and be highly selective to — the contaminant.

Finely ground activated carbon and charcoal have been used as adsorbents but they suffer from oxidation during thermal destruction of the contaminant. This makes them non-recyclable, thus both increasing cost and contributing CO_2 to the atmosphere. Smectite clay catalysts have been proposed as an alternative adsorbent, although some modifications of the natural mineral are necessary.

Modified smectite clay catalysts. Interlayer sites in smectite dehydrate at temperatures above 200°C, collapsing to an illitic structure. Since the ion exchange capacity of smectite centres on the interlayer site, collapse must be prevented if clay catalysts are to be used in thermal treatments of chemical organic toxins. The intercalation of thermally stable cations, which act as molecular props or pillars, is one method of keeping the interlayer sites open in the absence of a solvent like water (Fig. 3.20). Various pillaring agents can be used; the most common is the polynuclear hydroxyaluminium cation ($Al_{13}O_4(OH)_{28}^{3+}$), which is stable above 500°C. The provision of a pillar has two other advantages. First, it increases the internal surface area of the interlayer site, making it more effective as an adsorbent. Second, by introducing cationic props of different sizes and spacings (spacing is determined by the radius of the hydrated cation and the charge), it is possible to vary the size of spaces between props. It is thus possible to manufacture highly specific molecular sieves, which could be used to trap large ions or molecules (e.g. organic contaminants) whilst letting smaller, benign molecules pass through.

Smectite clays do not have a strong affinity for soluble organic contaminants. This is improved by using a surface-active agent (surfactant) (Box 3.11). Surfactant-coated interlayer sites provide a hydrophobic substrate (usually a long hydrocarbon chain) for which dioxins and other chlorinated phenols (also hydrophobic) have high affinity. The effect of adding the surfactant Tergitol 15s-5 to a clay catalyst is shown in Fig. 3.21.

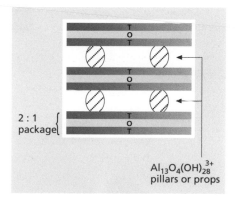

Fig. 3.20 Schematic diagram showing props or pillars in the interlayer position in smectite clay. T, tetrahedral sheet; O, octahedral sheet.

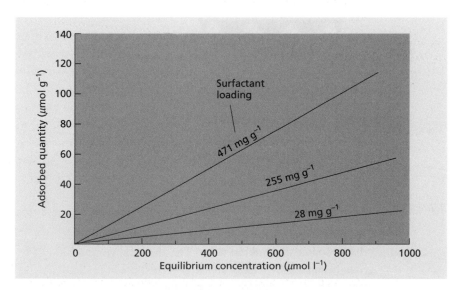

Fig. 3.21 Adsorption of 3-monochlorophenol from water on to Al_{13} smectite containing different amounts of Tergitol 15s-5 surfactant. From Michot & Pinnavaia (1991).

The modified smectite has good affinity and selectivity for its target contaminant and is thermally stable, recyclable and economic to use. A scheme summarising the use of modified smectite clay catalysts is sketched in Fig. 3.22.

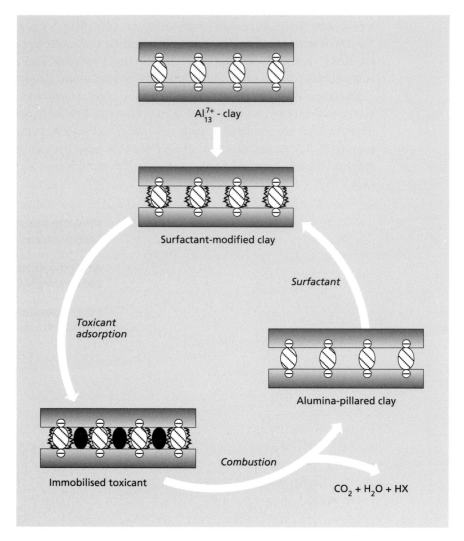

Fig. 3.22 Scheme for recycling surfactant-modified pillared clay mineral adsorbents during thermal treatment of toxicant. From Michot & Pinnavaia (1991).

3.7 The chemistry of continental waters

Continental freshwaters are very important to humans since they are the only reliable source of drinking-water. The chemical composition of rivers, lakes and groundwater varies widely and is governed predominantly by three factors: element chemistry, weathering regimes and biological processes. In addition, human perturbations may have a major effect on some freshwater systems.

Box 3.11

Surfactants

A surfactant (surface active agent) is a substance introduced into a liquid to affect (usually to increase) its spreading or wetting properties (i.e. those properties controlled by surface tension).

Detergents and soaps are examples of surfactants. A soap molecule has two features essential for its cleansing action: a long, non-polar hydrocarbon chain, and a polar group (carboxylate group). An example of a modern detergent, sodium lauryl sulphate, is shown below.

$$CH_3(CH_2)_{11}-O-\overset{\displaystyle O}{\underset{\displaystyle O}{\overset{\|}{\underset{\|}{S}}}}-O-Na^+$$

(non-polar hydrocarbon tail) (carboxylate head)

The polar carboxylate head dissolves in water, while the long hydrocarbon-chain tail mixes well with greasy substances, effectively floating the grease into solution.

In general terms, all surfactants behave like detergents, having a hydrophilic head and a hydrophobic tail. Consequently, a hydrophobic molecule or compound, for example, dioxins and other chlorinated phenols, will have affinity for the long hydrocarbon-chain tail.

3.7.1 Element chemistry

The 20 largest rivers on Earth carry about 40% of the total continental runoff, with the Amazon alone accounting for about 15% of the total. These rivers give the best indication of global average riverwater chemical composition, which can be compared with average continental crust composition (Table 3.8). Three features stand out from this comparison.

1 Four metals dominate the dissolved chemistry of freshwater, all present as simple cations (Ca^{2+}, Na^+, K^+ and Mg^{2+}).

2 The low concentration of ions in freshwater.

3 The dissolved ionic composition of freshwater is radically different from continental crust, despite the fact that all of the cations in riverwater, with the exception of some of the sodium and chloride (see Section 3.7.2), are derived from weathering processes.

The difference between crustal and dissolved riverwater composition is particularly marked for aluminium and iron relative to other metals (Table 3.8). This difference results from the way specific metal ions react with water.

Ionic compounds dissolve readily in polar solvents like water (see Box 3.1). Once in solution, however, different ions react with water in different ways. Low-charge ions (1+, 2+, 1–, 2–) usually dissolve as simple cations or anions. These ions have little interaction with the water, except that each ion is surrounded by water molecules (see Box 3.1). Smaller ions, with higher charge, react with water, abstracting OH^- to form uncharged and insoluble hydroxides, liberating hydrogen ions in the process, e.g.

$$Fe^{3+}_{(aq)} + 3H_2O_{(l)} \rightarrow Fe(OH)_{3(s)} + 3H^+_{(aq)} \qquad \text{eq. 3.21}$$

Still smaller and more highly charged ions react with water to produce relatively large and stable ions (so called oxyanions), such as sulphate (SO_4^{2-}), by abstracting oxygen ions from water and again liberating hydrogen ions, e.g.

$$S(VI)_{(s)} + 4H_2O_{(l)} \rightarrow SO_4^{2-}_{(aq)} + 2H^+_{(aq)} \qquad \text{eq. 3.22}$$

The net effect is to produce large anions, which dissolve readily since the charge is spread over a large ionic perimeter. Other important oxyanions are nitrate (NO_3^-) and carbonate (CO_3^{2-}).

Table 3.8 Comparison of the major cation composition of average continental crustal rock — from Taylor & McLennan (1985) — and average riverwater — from Berner & Berner (1987), except aluminium and iron from Broecker & Peng (1982)

	Continental crust (mg kg^{-1})	Riverwater (mg kg^{-1})
Al	80.0	0.05
Fe	35.0	0.04
Ca	30.0	13.4
Na	29.0	5.2
K	28.0	1.3
Mg	13.0	3.4

The general pattern of element solubility can be rationalised in terms of charge and ionic radius (z/r) (Fig. 3.23). Ions with low z/r values are highly soluble, form simple ions in solution and are enriched in the dissolved phase of riverwater compared with the particulate phase. Ions with intermediate z/r values are relatively insoluble and have a relatively high particulate : dissolved ratio in riverwater. Ions with large z/r values form complex oxyanions and again are soluble.

Some oxyanions exist in solution as weak acids such that their behaviour depends on solution pH, as shown for phosphorus.

$$H_3PO_{4(aq)} \rightleftharpoons H_2PO^-_{4(aq)} + H^+_{(aq)} \qquad \text{eq. 3.23}$$

$$H_2PO^-_{4(aq)} + H^+_{(aq)} \rightleftharpoons HPO^{2-}_{4(aq)} + 2H^+_{(aq)} \qquad \text{eq. 3.24}$$

$$HPO^{2-}_{4(aq)} + 2H^+_{(aq)} \rightleftharpoons PO^{3-}_{4(aq)} + 3H^+_{(aq)} \qquad \text{eq. 3.25}$$

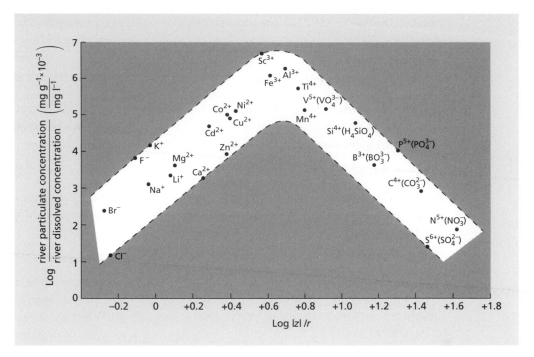

Fig. 3.23 Ratio of average elemental riverine particulate to dissolved concentrations plotted against the ratio of charge to ionic radius for the most abundant ions of those elements. In the case of dissolved oxyanions, the relevant dissolved species are shown in brackets. Concentration data from Martin & Whitfield (1983), other data from Krauskopf (1979).

3.7.2 Water chemistry and weathering regimes

Comparison of dissolved major ion compositions in four large rivers draining very different crustal areas (Table 3.9) shows the dominance of calcium, magnesium, sodium and potassium. Overall, however, the chemistry of each river is different and weathering regimes control most of these variations.

The dissolved ion composition of freshwater depends upon: the varying composition of rainfall and atmospheric dry deposition; the modification of atmospheric inputs by evapotranspiration; the varying inputs from weathering reactions and organic matter decomposition in soil and rocks; and differential uptake by biological processes in soils. Where crystalline rock or highly weathered tropical soils are present (i.e. where weathering inputs are low or exhausted), dissolved freshwater chemistry is most influenced by natural atmospheric inputs, e.g. sea spray and dust, as well as by anthropogenic gases, e.g. SO_2.

Sea-salt inputs, often called cyclic salts, are common in coastal regions. Small amounts of sea salts are, however, also present in rainwater of central continental areas, thousands of miles from the sea. Sea-salt inputs have broadly similar, predominantly sodium chloride (NaCl), chemistry to the seawater from which they were derived. Thus, sodium or chloride ions can be used as a measure of sea-salt inputs to rainwater.

Table 3.9 Dissolved major ion composition (mmol l⁻¹) of some major rivers. Data from Meybeck (1979), except Rio Grande which is from Livingston (1963)

	Mackenzie (1)	Orinoco (2)	Ganges (3)	Rio Grande (4)
Ca^{2+}	0.82	0.08	0.61	2.72
Mg^{2+}	0.43	0.04	0.20	0.99
Na^+	0.30	0.06	0.21	5.10
K^+	0.02	0.02	0.08	0.17
Cl^-	0.25	0.08	0.09	4.82
SO_4^{2-}	0.38	0.03	0.09	2.48
HCO_3^-	1.82	0.18	1.72	3.00
SiO_2	0.05	0.19	0.21	0.50

Drainage basin characteristics: (1) northern arctic Canada; (2) tropical northern South America; (3) southern Himalayas; (4) arid southwestern North America.

The importance of seawater sources for ions other than sodium and chloride in rainwater can be assessed by computing their relative abundance with respect to sodium and comparing this with the same ratio in seawater. This comparison can be extended to freshwater, although here there is the complication that some of the ions could be derived from weathering. If we overlook this complication initially, then, where rainwater inputs make a large contribution to the chemistry of freshwater, the dominant cation is likely to be Na^+. If weathering reactions are important, then the major dissolved cations will be those soluble elements derived from local rock and soil. In the absence of evaporite minerals, which are a minor component of continental crust (see Fig. 3.1), the most weatherable rocks are limestones ($CaCO_3$). The calcium ion, liberated by limestone dissolution, is an indicator of this weathering process. The ratio of $Na^+ : (Na^+ + Ca^{2+})$ can therefore be used to discriminate between rainwater and weathering sources in freshwaters. When sodium is the dominant cation (sea-salt contribution important), $Na^+ : (Na^+ + Ca^{2+})$ values approach 1. When calcium is the dominant ion (weathering contribution important), $Na^+ : (Na^+ + Ca^{2+})$ values approach 0.

The composition of dissolved ions in riverwater can be classified by comparing $Na^+ : (Na^+ + Ca^{2+})$ values with the total number of ions present in solution (Fig. 3.24). Data which plot in the bottom right of Fig. 3.24 represent rivers with low ion concentrations and sodium as the dominant cation. These rivers flow over crystalline bedrock (low weathering rates) or over extensively weathered, kaolinitic, tropical soils (low weathering potential, CIA c. 100 (see Table 3.6)). The Rio Negro, a tributary of the Amazon (Fig. 3.25), draining the highly weathered tropical soils of the central Amazonian region, has low ionic strength (Box 3.12) with sodium as the major cation. The Onyx River in the dry valleys of Antarctica is a better example of a low-ionic-strength, sodium-dominant river. This river forms from glacial melt water and begins with a chemistry almost totally dominated by marine ions. As it flows over the igneous and metamorphic rocks of the valley floor, its composition evolves to higher amounts of dissolved solids with an increasing proportion of calcium (see Fig. 3.24).

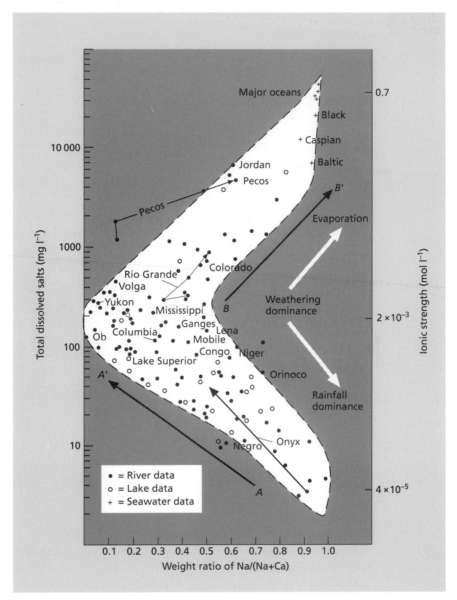

Fig. 3.24 Variation of the weight ratio of Na/(Na + Ca) as a function of the total dissolved solids and ionic strength for surface waters. Arrows represent chemical evolution of rivers from source downstream. *Science* **170**, 1088–1090, Gibbs, R.J. Copyright (1970) by the AAAS.

Major river systems flow over a wide range of rock types, acquiring the dissolved products of weathering reactions. Freshwaters originating in areas with active weathering processes will have higher ion concentrations and an increasing predominance of calcium over sodium. These rivers plot along a trend from A to A' on Fig. 3.24. The Mackenzie and Ganges (Table 3.9) fall within this group, despite very different geomorphological settings.

| Box 3.12 | |

Ionic strength

The concentration of an electrolyte solution can be expressed as ionic strength (I), defined as:

$$I = \tfrac{1}{2} \sum_i c_i z_i^2 \qquad \text{eq. 1}$$

where c_i is the concentration of ion (i) in mol l^{-1}, z_i is the charge of ion (i) and Σ represents the sum of all ions in the solution.

As a measure of the concentration of a complex electrolyte solution, ionic strength is better than simple sum of molar concentrations, as it accounts for the effect of charge of multivalent ions.

For example, the Onyx River in Antarctica (see Section 3.7.2), 2.5 km below its glacier source, has the following major ion composition (in μmol l^{-1}): Ca^{2+} = 55.4; Mg^{2+} = 44.4; Na^+ = 125; K^+ = 17.6; H^+ = 10^{-3}; Cl^- = 129; SO_4^{2-} = 32.2 and HCO_3^- = 136 OH^- = 10 (rivers in this region have pH around 9, which means that HCO_3^- is the dominant carbonate species). Putting these ions in eq. 1 gives:

$$I = \tfrac{1}{2} \Sigma \, c\mathrm{Ca}.4 + c\mathrm{Mg}.4 + c\mathrm{Na}.1 + c\mathrm{K}.1 + c\mathrm{H}.1 + c\mathrm{Cl}.1 + c\mathrm{SO}_4.4 + c\mathrm{HCO}_3.1$$
$$+ \, c\mathrm{OH}.1 \qquad \text{eq. 2}$$

Substituting the μmol l^{-1} values (and correcting to mol l^{-1} with the final 10^{-6} term in eq. 3) gives:

$$I = \tfrac{1}{2}[(55.4 \times 4) + (44.4 \times 4) + 125 + 17.6 + 10^{-3} + 129 + (32.2 \times 4) + 136$$
$$+ \, 10] \times 10^{-6} \qquad \text{eq. 3}$$

$$I = 4.73 \times 10^{-4} \text{ mol } l^{-1} \qquad \text{eq. 4}$$

Freshwaters typically have ionic strengths between 10^{-3} and 10^{-4} mol l^{-1}, whereas seawater has a fairly constant ionic strength of 0.7 mol l^{-1}.

The Amazon and its tributaries are a good example of a river system where the chemistry of the lower reaches integrates the products of differing soil and bedrock weathering regimes (Fig. 3.25). Rivers draining the intensely weathered soils and sediments of the central Amazonian region, such as the Rio Negro, have low total cation concentrations, < 200 μeq l^{-1} (i.e. sum of all major cations concentrations $\times$ charge; see also footnote to Table 3.7). The Rio Negro has water enriched in sodium, silica, iron, aluminium and hydrogen ions, because of the limited supply of other cations from weathering reactions. In contrast, rivers draining easily erodible sedimentary rocks (including carbonates) of the Peruvian Andes are characterised by high total cation concentrations of 450–3000 μeq l^{-1}, including abundant calcium, magnesium, alkalinity (see below and Box 3.13) and sulphate. Between these two extremes in water composition are rivers with quite low total cation concentrations (450–3000 μeq l^{-1}), with sodium enriched relative to calcium and magnesium, but also with high concentrations of silica, consistent with the weathering of feldspars (e.g. albite (see eq. 3.18)). These rivers drain areas without large amounts of easily weatherable rock, but drain soils not so completely degraded as the lowest concentration group characterised by the Rio Negro.

In arid areas, evaporation may influence the major dissolved ion chemistry of rivers. Evaporation concentrates the total amount of ions in riverwater. Evaporation also causes $CaCO_3$ to precipitate from water before NaCl, the latter being more soluble. The formation of $CaCO_3$ removes calcium ions from the water, increasing the $Na^+ : (Na^+ + Ca^{2+})$ value. Data for rivers influenced by evaporation plot along a diagonal B to B' on Fig. 3.24, evolving toward seawater composition (see Table 4.1). Examples include the rivers Jordan, Rio Grande and Colorado. As the waters of these rivers evolve downstream their compositional character, progressively altered by the effects of evaporation, may move toward the upper right of Fig. 3.24 (e.g. Rio Grande).

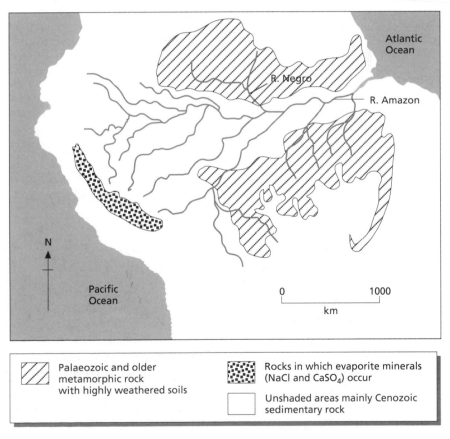

Fig. 3.25 Simplified geological map of the Amazon Basin showing major river systems. After Stallard & Edmond (1983).

The classification of riverwater composition in Fig. 3.24 is simplified and does not always work. For example, weathering of feldspars (see Section 3.4.4) can produce solutions of low ionic strength, but rich in sodium and silica, which plot in the bottom right of Fig. 3.24. This effect probably influences the classification of the Rio Negro. Weathering of evaporite minerals will also affect the composition of rivers. For example, in the Amazon catchment there are a small number of tributaries draining areas of predominantly evaporite rock. These have very high total cation concentrations and are characterised by high sodium, chloride, calcium and sulphate concentrations from the weathering of the evaporite minerals, halite and gypsum. Despite these complications, Fig. 3.24 remains a useful way to compare factors controlling riverwater chemistry. Indeed, it is remarkable that most of the world's major rivers can be rationalised in this straightforward way.

Finally, we should recall that most soilwaters that feed rivers and groundwater have near-neutral pH, with HCO_3^- as the major anion. This results from the dissolution of CO_2 in water and from the acid hydrolysis of silicates and carbonates (see Section 3.4.3). The total amount of weak anions in water is often referred to as alkalinity, and the anions maintain (buffer) the pH around 8 (Box 3.13).

Box 3.13

Alkalinity and pH buffering

Alkalinity is a measure of the total amount of weak acid anions in a solution which are available to neutralise H^+. In natural waters, bicarbonate (HCO_3^-) and carbonate (CO_3^{2-}) ions are most important, although in seawater other ions also contribute to alkalinity. The relative importance of HCO_3^- and CO_3^{2-} depends on the pH of the solution and can be calculated from the known dissociation constants of these ions and the solution pH. The alkalinity is measured by adding acid to a water sample until the pH falls to 4. At this pH, HCO_3^- and CO_3^{2-} alkalinity will have been converted to carbon dioxide (CO_2), i.e.:

$$HCO_{3(aq)}^- + H_{(aq)}^+ \rightleftharpoons H_2O_{(l)} + CO_{2(g)} \qquad \text{eq. 1}$$

$$CO_{3(aq)}^{2-} + 2H_{(aq)}^+ \rightleftharpoons H_2O_{(l)} + CO_{2(g)} \qquad \text{eq. 2}$$

The volume of acid used is a measure of the alkalinity, which is usually expressed as milliequivalents per litre (see footnote to Table 3.7).

In natural waters, pH is controlled mainly by the concentration of dissolved CO_2, HCO_3 and CO_3^{2-}. These species react to maintain the pH within relatively narrow limits. This is known as buffering the pH. The principles of pH buffering, using worked examples, are given in Appendix 2, pp. 200–202.

3.7.3 Silicon and aluminium

Silicon is mobilised by the weathering of silicate minerals and is transported in natural waters as undissociated silicic acid, H_4SiO_4. Silicate minerals weather slowly, such that rates of input — and concentrations — of silicon in most freshwaters are quite low. Despite this, where silicates are the main component of bedrock or soil, H_4SiO_4 can be a significant component of the total dissolved solids in freshwater.

Aluminium is largely insoluble during weathering processes (see Table 3.8), but becomes soluble when pH is both low and high. At the simplest level, three aluminium species are identified; soluble Al^{3+}, dominant under acid conditions, insoluble aluminium hydroxide $(Al(OH)_3)$, dominant under neutral conditions, and $Al(OH)_4^-$, dominant under alkaline conditions.

$$Al(OH)_{3(s)} + OH^-_{(aq)} \rightleftharpoons Al(OH)_4^- \qquad \text{eq. 3.26}$$

$$Al(OH)_{3(s)} \rightleftharpoons Al^{3+}_{(aq)} + 3OH^-_{(aq)} \qquad \text{eq. 3.27}$$

Aluminium solubility is therefore pH-dependent and aluminium is insoluble in the pH range 5–9, which includes most natural waters. The details of aluminium solubility are complicated by the formation of partially dissociated $Al(OH)_3$ species and complexing between aluminium and organic matter. An understanding of the controls on aluminium solubility are important since aluminium toxicity may cause fish deaths in acidified freshwaters.

Acidification of soilwater occurs if the rate of displacement of soil cations by H^+ exceeds the rate of cation supply from weathering. Ion exchange reactions (reverse of eq. 3.20) help to buffer pH in the short term, but over longer periods cation supply to soils is from the underlying bedrock. Rainwater is naturally acidic (see Box 2.12) and soilwaters are further acidified by the production of H^+ from the decomposition of organic matter (eqs 3.10–3.14). Thus, acidification can be a natural process, although acid rain (see Section 2.9) has greatly increased the rate of these processes in many areas of the world.

Acidification of freshwater is most marked in upland areas with high rainfall (hence high acid flux), steep slopes (resulting in a short residence time for water in the soil) and crystalline rocks, which weather — and supply cations — slowly. Thus, while acid rain is a widespread phenomenon, acidified freshwaters are less common and are controlled both by rates of atmospheric input and by rock types (Fig. 3.26). All weathering processes, except sulphide oxidation (see Section 3.4.2), consume hydrogen ions, driving pH toward neutrality. Hence, mature rivers, which drain deeper, cation-rich lowland soils, have higher pH and lower aluminium concentrations.

The effects of upland acidification of freshwaters can be dramatic. Between 1930 and 1975 the median pH of lakes in the Adirondack Mountains of northeastern USA decreased from 6.7 to 5.1, caused by progressively lower pH in rainwater (Fig. 3.26). The acidified lakewater killed fish and other animals. Similar problems have been reported from Scandinavia and Scotland. In addition to problems in freshwaters, the loss of forests in high-altitude areas has been linked to acid leaching, which leads to impoverishment of soils coupled with direct loss of cations from plant leaves.

Although aluminium is soluble at high pH, alkaline waters are uncommon because they absorb acid gases, e.g. CO_2 and SO_2, from the atmosphere. However,

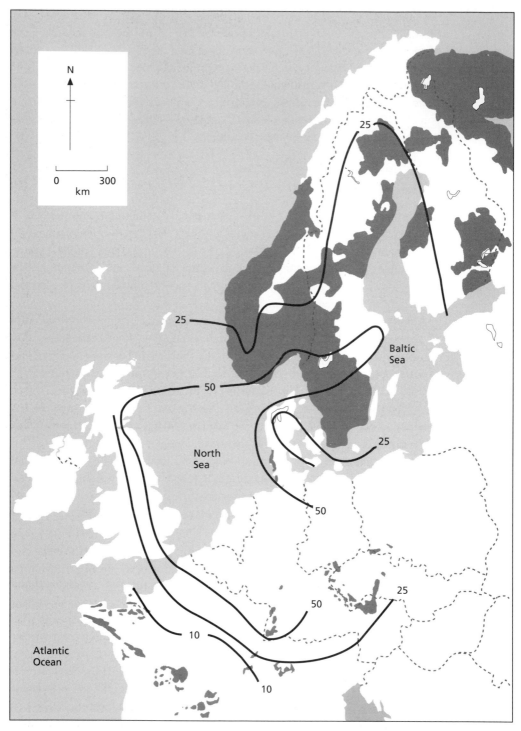

Fig. 3.26 Rates of acid deposition (contours of μmol H^+ l^{-1}) and areas most sensitive to acidification (shaded) based on their rock type: (a) Europe.

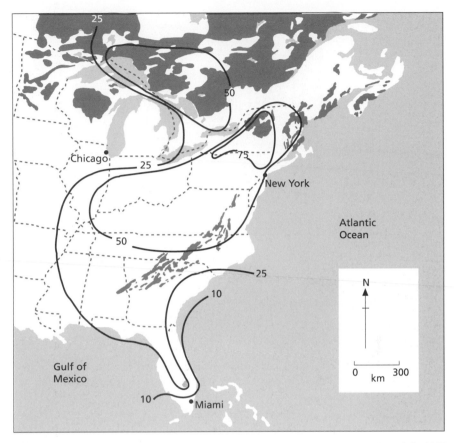

Fig. 3.26 (*Continued*) (b) North America. From *Acid Rain*, Likens, G.E. *et al.* Copyright 1979 by Scientific American, Inc. All rights reserved.

alkaline rivers with aluminium mobility are known. The industrial process for abstracting aluminium from bauxite involves leaching the ore with strong sodium hydroxide (NaOH) solutions. In Jamaica, discharge of wastes from bauxite processing produces freshwater streams with high pH in addition to high sodium and aluminium concentrations. As these streamwaters evolve, the pH falls to about 8 and the dissolved aluminium : sodium ratio declines as the aluminium precipitates (Fig. 3.27). The importance of pH with regard to water chemistry is demonstrated equally well by groundwater contamination problems (see Section 3.7.6).

3.7.4   Biological processes

In streams and small rivers, biological activity in the water has little influence on water chemistry because any effects are diluted by the rapid flow. Conversely, in large slow-flowing rivers and in lakes, biological activity can cause major changes in water chemistry.

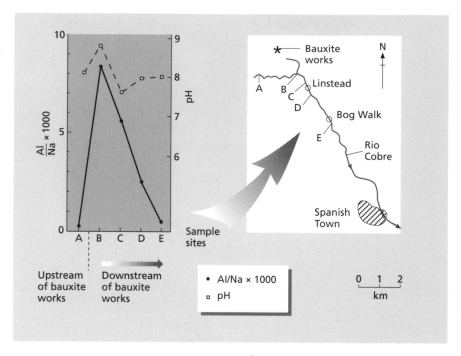

Fig. 3.27 Sampling sites on the Rio Cobre (southern Jamaica) and ratio of dissolved aluminium to sodium concentrations, and pH, at those sites. Unpublished data courtesy of T.D. Jickells and A. Greenaway.

All photosynthetic plants absorb light and convert this to chemical energy within a chlorophyll molecule. The liberated energy is then used to convert CO_2 (or HCO_3^-) and water into organic matter in the following way.

$$CO_{2(g)} + H_2O_{(l)} \xrightarrow{\text{light}} CH_2O_{(s)} + O_{2(g)} \qquad \text{eq. 3.28}$$

As in eq. 3.10, CH_2O represents the generalised formula for carbohydrate organic matter. The reaction depicted by eq. 3.28 requires the input of energy ($\Delta G° = +475$ kJ mol^{-1}) (see Box 3.7) to proceed and this is provided by light. In shallow freshwater, large plants and drifting microscopic algae (phytoplankton) are responsible for photosynthesis, while in deep lakes (and in the oceans) phytoplankton account for almost all photosynthesis. The reverse reaction of organic matter decomposition, i.e. oxidation or respiration, liberates the energy which sustains all life.

$$CH_2O_{(s)} + O_{2(g)} \rightarrow CO_{2(g)} + H_2O_{(l)} \qquad \Delta G° = -475 \text{ kJ mol}^{-1} \qquad \text{eq. 3.29}$$

Since photosynthesis requires light, it is confined to the surface layers of waters — the euphotic zone (the region receiving >1% of the irradiance arriving at the water surface). The depth of the euphotic zone varies with the angle of the sun, the amount of light absorbed by suspended matter (including phytoplankton) and the presence of dissolved coloured compounds in the water.

The decomposition of organic matter (which is almost always bacterially mediated) can occur at any depth in the water column. Decomposition consumes oxygen (eq. 3.29), which is supplied to the water largely by gas exchange at the water/air interface and partly as a by-product of photosynthesis. Temperature influences the amount of oxygen that can dissolve in water. Oxygen-saturated freshwater holds about 450 μmol l^{-1} oxygen at 1°C and 280 μmol l^{-1} at 20°C.

In summer, the surface layers of many lakes are warmed by insolation. The warmer surface water is less dense than the cold deep water, causing a stable density stratification. Stratification limits exchange of oxygenated surface water with the deeper waters. Organic matter, produced in surface waters, sinks into the deeper waters, where it is oxidised, further depleting oxygen concentrations. In some cases, oxygen levels fall below those needed to support animal life. The rate of oxygen consumption increases as the supply of organic matter increases, either due to enhanced photosynthesis in the surface waters, or due to direct discharge of organic waste, e.g. sewage. The Thames estuary, Chesapeake Bay (see Section 4.2.4) and the Baltic Sea (see Box 4.9) are examples of water bodies affected by low oxygen concentrations, whilst the specialised case of stratification in an African lake is discussed in Box 3.14.

Once oxygen has been used up, bacteria use alternative oxidising agents to consume organic matter (see Box 3.5). These alternative oxidants are used in an order which depends on energy yields (Table 3.10). Nitrate reduction (denitrification) is energetically favourable to bacteria, but is often limited in natural freshwaters by low nitrate concentrations. Anthropogenic inputs, however, have resulted in increased nitrate concentrations in rivers and groundwater (see Section 3.7.5), increasing the availability of nitrate for bacterial reduction.

Iron and manganese (Mn), both potential electron acceptors, are common as insoluble Fe(III) and Mn(IV) oxides. In reducing environments (at about the same redox potential as nitrate reduction), these oxides may be reduced to soluble Fe(II) and Mn(II) (Table 3.10). Indeed, iron is soluble only under low redox or acidic conditions (Box 3.15).

Sulphate reduction (Table 3.10) is not an important mechanism of organic matter respiration in freshwaters because dissolved sulphate levels are usually low. In seawater, however, sulphate is abundant and sulphate reduction is very important (see Section 4.4.6). Methanogenesis can be an important respiration process in some organic-rich freshwater lake and swamp sediments. The reduced reaction product, methane (CH_4), a greenhouse gas (see Section 5.3.4), is known to bubble out of some wetlands, including rice paddy fields, contributing significantly to atmospheric CH_4 budgets (see Section 2.3).

3.7.5 Nutrients and eutrophication

In addition to CO_2, water and light, ions (or nutrients) are needed for plant growth. Some of these ions, e.g. Mg^{2+}, are abundant in freshwater, but other essential nutrients, e.g. nitrogen (N) and phosphorus (P), are present at low concentrations. If

Box 3.14

The Lake Nyos gas disaster: a natural hazard related to lake stratification

In sheltered locations, deep lakes stratify due to density differences. These are caused by gases or salts dissolved in the waters, or surface warming, which keeps buoyant water layers floating on deep, dense layers (the hypolimnion) (Fig. 1). At times this stratification may break down naturally, allowing layers to overturn and mix. In temperate regions, overturn is common in the autumn when surface layers cool and sink, displacing the bottom water. Seasonal overturn in the tropics is not common due to the minor seasonal weather changes. However, Lake Barombi Mbo in Cameroon, West Africa, is known to overturn in late August and September, probably because of the cloudiness of the monsoon, which reduces solar insolation on the surface waters.

Lake Nyos, a volcanic crater lake in Cameroon, is also density-stratified, caused largely by increased dissolved carbon dioxide (CO_2) content of the deep water. Over time, CO_2 degassing from a magma at shallow depth in the crust seems to leak steadily into the bottom waters, building up to saturated concentrations.

On 21 August 1986, approximately 1×10^9 m³ of CO_2 was released from the bottom water as an enormous cloud. Carbon dioxide, being 1.5 times heavier than air, flowed into nearby valleys, asphyxiating thousands of people. The reason for this disastrous gas burst is not known. It may be that a local earthquake or the accumulation and sinking of cold rainwater runoff following heavy rain caused rapid overturn, releasing CO_2 from the bottom waters (Fig. 2). Alternatively, it may be that degassing occurred without complete overturn of the lake. If local CO_2 supersaturation led to spontaneous CO_2 gas bubble formation, then local decompression might have initiated a chain reaction, leading to further CO_2 degassing. Any small upward movement of deeper lake waters, caused perhaps by wave disturbance, could have set off this chain reaction (Fig. 3).

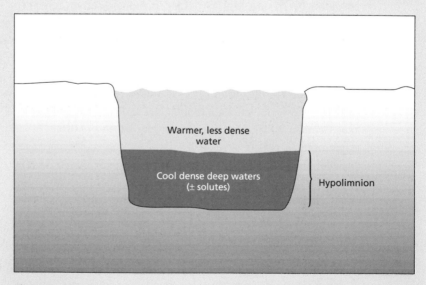

Fig. 1 Stable stratified condition.

Box 3.14
Cont.

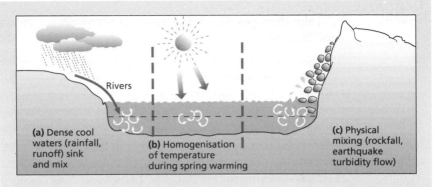

Fig. 2 Alternative mechanisms to cause overturn.

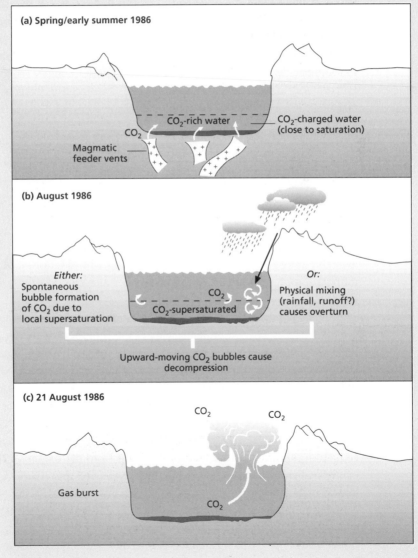

Fig. 3 Lake Nyos 1986.

light availability does not limit algal growth, chemical limitation is likely to occur when demand for nitrogen and phosphorus exceeds their availability. Consequently, a great deal of attention has been focused on the behaviour of nitrogen and phosphorus in natural waters and their role as potential, or actual, limiting nutrients. In seawater, the ratio of nitrogen to phosphorus required for optimal growth is quite well known, being 16 : 1 on an atomic basis. In freshwater, the required nitrogen : phosphorus ratio is more variable.

In natural waters, dissolved inorganic phosphorus (DIP) exists predominantly as various dissociation products of phosphoric acid (H_3PO_4) (see eqs 3.23–3.25). Phosphorus is usually retained in soils by the precipitation of insoluble calcium and iron phosphates, by adsorption on iron hydroxides or by adsorption on to soil particles. As a result, DIP in rivers is derived mainly from direct discharges, e.g. sewage. DIP concentrations vary inversely with river flow (Fig. 3.28), the input being diluted under higher flow conditions. Since phosphate is usually in sediments as insoluble iron(III) phosphate ($FePO_4$), under reducing conditions (such as occur in sediments when oxygen consumption exceeds supply), DIP can be returned to the water column in association with iron(III) reduction to iron(II).

Nitrogen chemistry is complex because nitrogen can exist in several oxidation states, of which N(0) nitrogen gas (N_2), N(3–) ammonium (NH_4^+) and N(5+) nitrate (NO_3^-) are the most important. Nitrogen gas dissolved in riverwater cannot be utilised as a nitrogen source by most plants and algae because they cannot break its strong triple bond (see Box 2.2). Specialised 'nitrogen-fixing' bacteria and fungi do exist to exploit N_2, but it is not an energetically efficient way of obtaining nitrogen. Hence, these microorganisms are only abundant when N_2 is the only available nitrogen source. Nevertheless, along with fixation of N_2 by lightning, nitrogen-fixing microorganisms provide the major natural source of nitrogen for rivers.

Biological processes use nitrogen in the 3– oxidation state, particularly as amine groups (see Box 2.7) in proteins. This is the preferred oxidation state for algal uptake and also the form in which nitrogen is released during organic matter decomposition, largely as NH_4^+. Once released into soils or water, NH_4^+, being cationic, may be adsorbed on to negatively charged organic coatings on soil particles or clay mineral surfaces. Ammonium is also taken up by plants or algae, or oxidised to NO_3^-, a process that is usually catalysed by bacteria.

In contrast to NH_4^+, NO_3^- is anionic, soluble and not retained in soils. Therefore, NO_3^- from rainwater or fertilisers, or derived from the oxidation of soil organic matter and animal wastes, will wash out of soils and into rivers. Apart from biological uptake, denitrification in low-oxygen environments is the most important way that nitrate is removed from soils, rivers and groundwater. It has been estimated that, in the rivers of northwest Europe, half of the total nitrogen input to the catchment is lost by denitrification before the waters reach the sea. Thus, under low redox conditions, DIP is mobilised during iron(III) reduction and NO_3^- is lost, again emphasising the importance of redox processes in environmental chemistry.

Table 3.10 Order of bacterial reactions during microbial respiration of organic matter based on energy yield. After Berner, R.A. *Early Diagenesis.* Copyright © 1980 by PUP. Reproduced by permission of Princeton University Press

	ΔG° (kJ mol^{-1} of CH_2O)
Aerobic respiration	
$CH_2O + O_2 \rightarrow CO_2 + H_2O$	-475
Denitrification	
$5CH_2O + 4NO_3^- \rightarrow 2N_2 + 4HCO_3^- + CO_2 + 3H_2O$	-448
Manganese reduction	
$CH_2O + 3CO_2 + H_2O + 2MnO_2 \rightarrow 2Mn^{2+} + 4HCO_3^-$	-349
Iron reduction	
$CH_2O + 7CO_2 + 4Fe(OH)_3 \rightarrow 4Fe^{2+} + 8HCO_3^- + 3H_2O$	-114
Sulphate reduction	
$2CH_2O + SO_4^{2-} \rightarrow H_2S + 2HCO_3^-$	-77
Methanogenesis	
$2CH_2O \rightarrow CH_4 + CO_2$	-58

Free energy value for organic matter (CH_2O) is that of sucrose.

The very different chemistry of DIP and NO_3^- is illustrated by their behaviour in groundwater. On the limestone island of Bermuda there is little surface water because rainfall drains rapidly through the permeable rock to form groundwater. Almost all of the sewage waste on Bermuda is discharged to porous walled pits which allow effluent to gradually seep into the groundwater. The sewage has a nitrogen : phosphorus ratio of about 16 : 1, and yet Bermudian groundwater is characterised by very low DIP concentrations (average 3.5 µmol l^{-1}) and very high NO_3^- concentrations (average 750 µmol l^{-1}). The nitrogen : phosphorus ratio of 215 : 1 for the groundwater implies removal of $> 90\%$ of the DIP. High nitrate concentrations are characteristic of groundwater in areas of intensive agriculture and can compromise its use as a drinking-water supply (see Section 3.7.6). To safeguard drinking-water supplies in southeast England, farmers are required to control fertiliser inputs in areas of groundwater recharge.

The seasonal variation of NO_3^- concentrations in many temperate rivers is caused by fluctuation in supply of NO_3^- from soil. In summer, NO_3^- concentrations are low because soilwater flushing by rainfall is low. In the autumn, soil moisture content increases, allowing nitrate to wash out of the soil into rivers (Fig. 3.29). Increases

Box 3.15

Eh–pH diagrams

Acidity (pH) and redox potential (Eh) may determine the chemical behaviour of elements or compounds in an environment. In theory, an infinite variey of Eh–pH combinations is possible, although the pH of most environments on Earth is between 0 and 14 and more usually between 3 and 10. Redox potential is constrained by the existence of water. Under very oxidising conditions (Eh 0.6–1.2 V) water is broken into oxygen and hydrogen ions and under highly reducing conditions (Eh 0.0 to –0.6 V) water is reduced to hydrogen. Eh–pH diagrams are used to visualise the effects of changing acidity or redox conditions. The

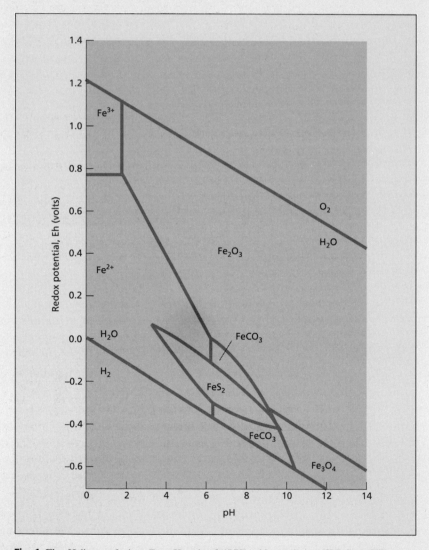

Fig. 1 Eh–pH diagram for iron. From Krauskopf (1979), with permission of McGraw-Hill.

**Box 3.15
Cont.**

diagram for iron is typical (Fig. 1). The lines represent conditions under which species on either side are present in equal concentrations. The exact position of the lines varies depending on the activities of the various species.

From the diagram it is clear that haematite (Fe_2O_3) is usually the stable iron species under oxidising conditions with pH above 4. Soluble Fe^{3+} is only present under very acidic conditions because of its tendency to form insoluble hydroxides (see Section 3.7.1). This tendency is only overcome under very acid conditions when hydroxide ion (OH^-) concentrations are low. Fe^{2+} is less prone to form insoluble hydroxides because of its small z/r value (see Section 3.7.1). Fe^{2+} is thus soluble at higher pH, but can only persist under low Eh conditions, which prevent its oxidation to Fe^{3+}. The small stability field for the common sulphide, pyrite (FeS_2), shows that this mineral can only form under reducing conditions, usually between pH 6 and 8.

in both the area and the intensity of agricultural activity are probably responsible for the increased NO_3^- concentrations seen in British (Fig. 3.30), other European and North American rivers.

Artificially increased NO_3^- concentrations mean that DIP is now the main limiting nutrient for plant growth in many freshwaters. Increased algal biomass can cause toxicity, clogging of water filters, unsightly water bodies, reduced biodiversity and low oxygen concentrations in stratified waters, a process usually referred to as eutrophication. The relationship between DIP and chlorophyll levels (a measure of algal biomass) (Fig. 3.31) makes the management of phosphorus inputs to rivers and lakes very important.

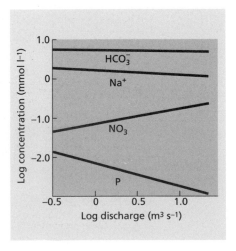

Fig. 3.28 Relationship between dissolved ion concentration and river discharge in the River Yare (Norfolk, UK). After Edwards (1973). Direct discharges of phosphorus from sewage and products of weathering (HCO_3^- and Na^+) decline in concentration (are diluted) as discharge increases. In contrast, heavy rainfall leaches NO_3^- from soil, causing NO_3^- concentrations to rise as discharge increases.

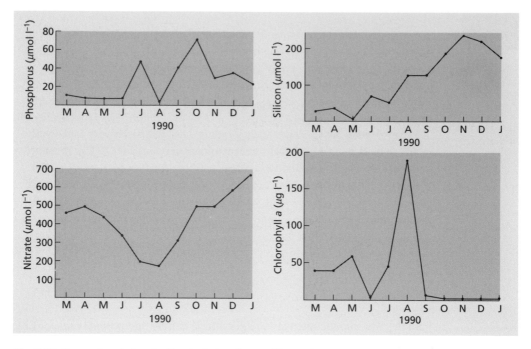

Fig. 3.29 Seasonal variations in dissolved phosphorus, silicon, nitrate and particulate chlorophyll *a* (as a measure of phytoplankton abundance) in the River Great Ouse (eastern England). Data from Fichez *et al.* (1992).

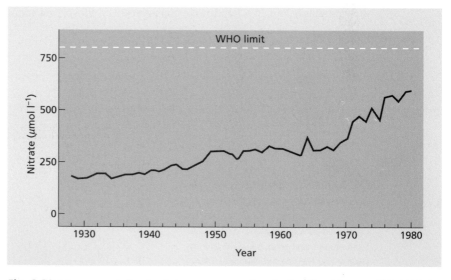

Fig. 3.30 Mean annual dissolved nitrate concentrations in the River Thames, UK, 1930–80. WHO limit refers to the World Health Organisation recommended maximum safe nitrate concentration in drinking-water. After Greenwood (1982).

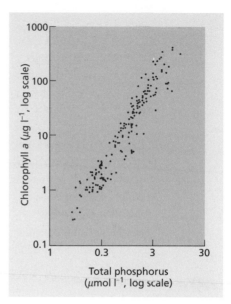

Fig. 3.31 Relationship between summer particulate chlorophyll *a* levels (as a measure of phytoplankton abundance) and total phosphorus concentrations in various lakes, with data plotted on log scales. After Moss (1988).

One other important nutrient, silicon, is used by diatoms (a group of phytoplankton) to build their exoskeletons. Diatoms are capable of rapid and prolific growth in nutrient-rich conditions. In temperate rivers, diatom blooms occur early in the year. For example, in the River Great Ouse in eastern England, silicon levels fall in early spring as diatom growth begins and rise again in summer as diatoms are displaced by other algal groups (see Fig. 3.29). In nitrate-rich rivers like the Great Ouse (nitrogen : phosphorus about 30 : 1 in winter), biological production makes little initial impact on NO_3^- levels until later in the season as the NO_3^- supply decreases due to reduced runoff. A NO_3^- minimum is reached in summer, before rising again in the autumn (see Fig. 3.29). DIP concentrations, in contrast, show more erratic behaviour (see Fig. 3.29), reflecting the influences of biological and dilution control, but are generally higher during low-flow conditions in summer. Since silicon is derived entirely from weathering reactions, its naturally low concentrations may be drastically reduced by diatom blooms, such that further diatom growth is limited. Thus silicon availability limits species diversity but not total phytoplankton biomass.

3.7.6 Contamination of groundwater

Groundwater is critically important to humans since it is a major source of drinking-water. For example, in the USA over 50% of the population rely on groundwater as a source of drinking-water. Groundwater quality is therefore very important and, in most developed countries, water must conform to certain standards for human consumption. Groundwater may fail to meet water quality standards because it contains dissolved constituents arising from either natural or anthropogenic sources. Typical anthropogenic mechanisms of groundwater contamination are given in Fig. 3.32. In the USA, major threats to groundwater include spillage from underground storage tanks, effluent from septic tanks and leachate from agricultural activities, municipal landfills and abandoned hazardous waste sites. The most frequently reported contaminants from these sources include nitrates, pesticides, volatile organic compounds, petroleum products, metals and synthetic organic chemicals.

The chemistry of contaminated groundwater is little different from that of surface waters. However, degradation processes that occur in days or weeks in surface waters may take decades in groundwater, where flow rates are slow and microbiological activity is low. This limits the potential for natural purification through flushing or biological consumption. Once contaminated, groundwater is difficult and expensive — in many cases impossible — to rehabilitate. The location of older sites of contamination may be imprecisely known, or even unknown, and hydrogeological conditions may dictate that contaminated groundwater discharges at natural springs into rivers or lakes, spreading contamination to surface waters.

The following case-studies highlight different styles of groundwater contamination where chemical considerations have proved important.

Case-study 1: Landfill leakage — Babylon, Long Island, New York, USA Shallow groundwater contamination of a surface sand aquifer has resulted from leakage of leachate rich in Cl^-, nitrogen compounds, trace metals and a complex mixture of organic compounds from Babylon landfill site, New York. Landfilling began in the 1940s with urban and industrial refuse and cesspool waste. The refuse layer is now about 20 m thick, some of it lying below the water-table. Chloride behaves conservatively (see Section 4.2.2) and is thus an excellent tracer of the contaminant plume, which is now about 3 km long (Fig. 3.33).

Close to the landfill, most nitrogen species are present as NH_4^+, indicating reducing conditions resulting from microbial decomposition of organic wastes. With increasing distance from the landfill, NO_3^- becomes quantitatively important due to the oxidation of NH_4^+, brought about by mixing of the leachate plume with oxygenated groundwater. This demonstrates how nitrogen speciation can be used to assess redox conditions in a contaminant plume.

Reducing conditions within the leachate plume also cause metal mobility, particularly of manganese and iron. The plume near the landfill has a pH of 6.0–6.5 and is reducing (–50 mV), making Fe^{2+} stable (Box 3.15). The transition to oxidising conditions downgradient in the aquifer allows solid iron oxides (e.g. FeOOH) to precipitate, dramatically reducing the mobility of metals which coprecipitate with iron.

Fig. 3.32 (*Facing page*) Typical anthropogenic mechanisms of groundwater contamination.

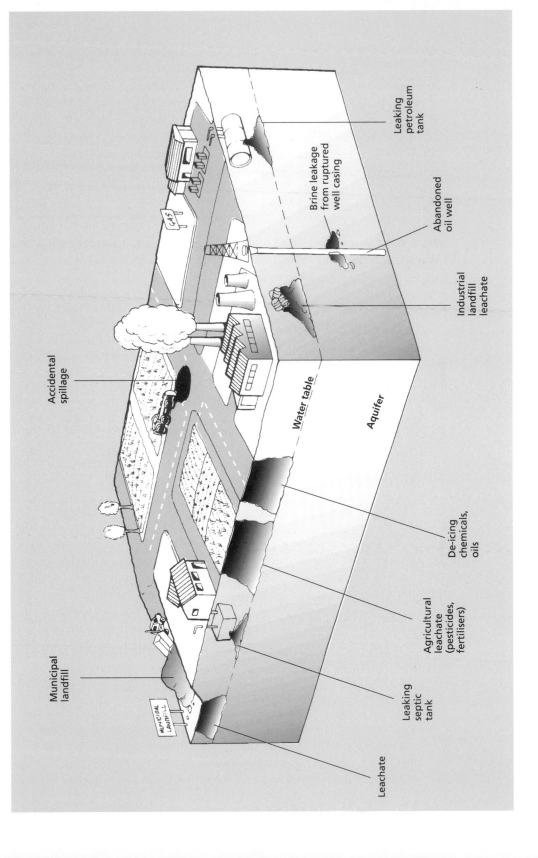

Leaking petroleum tank

Brine leakage from ruptured well casing

Abandoned oil well

Industrial landfill leachate

Accidental spillage

Water table

Aquifer

De-icing chemicals, oils

Agricultural leachate (pesticides, fertilisers)

Leaking septic tank

Municipal landfill

Leachate

GAS

MUNICIPAL LANDFILL

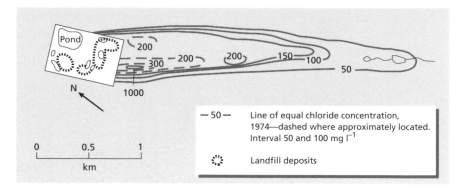

Fig. 3.33 Map of Cl⁻ plume at 9–12 m depth below the water-table, Babylon landfill site, 1974, showing the extent of groundwater contamination. After Kimmel & Braids (1980).

This relatively inoffensive example illustrates the importance of redox conditions in contaminated groundwater. Worse scenarios are known where chlorophenolic compounds, in very alkaline groundwaters (pH 10), ionise to negatively charged species and become much more mobile than in the neutral conditions generally typical of groundwater.

Case-study 2: Bowling Green, Kentucky, USA. In karst regions, limestone bedrock is heavily fissured and joints in the rock are enlarged by dissolution, resulting in interconnected caves. Sinkholes may divert surface streams into these fissures and caves, resulting in a subsurface drainage system. Accidental spillage of toxic chemicals or any other contaminant are rapidly dispersed in these conduits, making remediation particularly difficult.

The city of Bowling Green, Kentucky, is built on the karstified Ste Geneviève limestone, with underground drainage through the Lost River Cave (as shown in Fig. 3.34). It is the volatility (i.e. the tendency to transform into the gas phase) of certain chemical compounds that has caused specific problems at Bowling Green.

In recent decades, individual leakages of petroleum, up to 22 000 l volume, from storage tanks at auto service stations, have allowed petroleum to percolate into the subsurface water. Petroleum floats on the surface of groundwater within the karst system, rapidly filling caves with explosive fumes, particularly at sumps or traps in the cave system (Fig. 3.34). The trapped fumes have then risen into dwelling basements, water wells and storm drains.

In addition, leaking tanks at a chemical company are believed to have delivered benzene (C_6H_6), methylene chloride (CH_2Cl_2), toluene ($C_6H_5CH_3$), xylene ($C_6H_4(CH_3)_2$) and aliphatic hydrocarbons (i.e. hydrocarbons composed of open chains of carbon atoms) to subsurface water. These toxic (some carcinogenic) chemicals vaporise in the cave atmosphere, collect at traps and then rise into homes in a similar way to petroleum fumes.

The potential explosive/toxicity risk in Bowling Green has resulted in a number

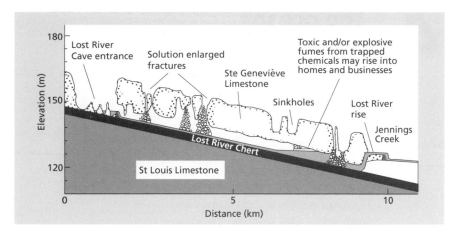

Fig. 3.34 Cross-section through the Lost River Cave drainage system underlying the city of Bowling Green, Kentucky, showing potential trap for floating or gaseous contaminants. Reprinted from: Beck, B.F. (ed), *Sinkholes: Their Geology, Engineering and Environmental Impact* — First multidisciplinary conference on sinkholes, Orlando, Florida, 15–17 October 1984. A.A. Balkema. P.O. Box 1675, Rotterdam, Netherlands.

of evacuations of homes in the last 20 years. Remediation measures have included better storage tank containment, regular monitoring of cave conduit outlets and ventilation of basements in homes at risk. It is hoped that these will prevent a disaster such as occurred in nearby Louisville, Kentucky, where an underground sewer explosion travelled 11 blocks, causing over 43 million dollars' worth of damage.

3.8   Further reading

Berner, K.B. & Berner, R.A. (1987) *The Global Water Cycle*. Prentice Hall, Englewood Cliffs, NJ, 397 pp.

Birkeland, P.W. (1974) *Pedology, Weathering, and Geomorphological Research*. Oxford University Press, New York, 285 pp.

Gill, R. (1989) *Chemical Fundamentals of Geology*. Unwin Hyman, London, 292 pp.

Harrison, R.M., deMora, S.J., Rapsomanikas, S. & Johnston, W.R. (1991) *Introductory Chemistry for the Environmental Sciences*. Cambridge University Press, Cambridge, 354 pp.

Moss, B. (1988) *Ecology of Freshwaters: Man and Medium*, 2nd edn. Blackwell Scientific Publications, Oxford, 417 pp.

Stumm, W. & Morgan, J.J. (1981) *Aquatic Chemistry*. Wiley, New York, 780 pp.

Veizer, J. (1988) The evolving exogenetic cycle. In: *Chemical Cycles in the Evolution of the Earth*, ed. by Gregor, C., Garrels, R.M., Mackenzie, F.T. & Maynard, J.B., pp. 175–220. Wiley, New York.

4 The oceans

4.1   Introduction

The oceans are by far the largest reservoir of the hydrosphere (see Fig. 1.4) and have existed for at least 3.8 billion years. Life on Earth probably began in seawater and the oceans are important in moderating global temperature changes. Riverwater, draining continental land areas, enters the oceans through estuaries. Here, freshwater mixes with seawater. The chemical composition of seawater is quite different from that of freshwater, a difference which affects the transport of some dissolved and particulate components. In addition, humans often perturb the natural chemistry of coastal areas, either through contamination of the freshwater runoff, or due to activities located close to estuaries and shallow seas.

We will begin by examining the chemistry of seawater close to continental areas, in the transition zone between terrestrial and open-ocean environments.

4.2   Estuarine processes

There are many differences between the chemistry of continental surface waters and seawater. In particular, seawater has a much higher ionic strength than most continental water (see Fig. 3.24) and seawater has a huge concentration of sodium and chloride ions (Na^+ and Cl^-) (Table 4.1), in contrast with calcium bicarbonate-dominated continental waters (see Section 3.7.2). Seawater is a strong chemical solution, such that mixing only 1% (volume) of seawater with average riverwater produces a solution in which the ratio of most ions, one to another, is almost the same as in seawater. Thus, the chemical gradients in estuaries are very steep and localised to the earliest stages of mixing. In addition to the steep gradient in ionic strength, in some estuaries there is also a gradient in pH.

Unidirectional flow in rivers is replaced by tidal (reversing) flows in estuaries. At high and low tide, water velocity drops to zero, allowing up to 95% of fine-grained suspended sediment (mainly clay minerals and organic matter) to sink and deposit. The efficiency of estuaries as sediment traps has probably varied over quite short geological timescales. For example, over the last 11 000 years as sealevel has risen following the last glaciation, estuaries seem to have been filling with sediment reworked from continental shelves. We might regard estuaries as temporary features on a geological timescale but this does not reduce their importance as traps for riverine particulate matter today.

4.2.1   Aggregation of colloidal material in estuaries

In estuarine water the steep gradient in ionic strength destabilises colloidal material (i.e. a suspension of fine-grained material), causing it to flocculate and sink to the bed. We can better understand this by considering clay minerals, the most abundant inorganic colloids in estuarine waters. Clay minerals have a surface negative charge (see Section 3.6.6) that is partly balanced by adsorbed cations. If surface charges are not neutralised by ion adsorption, clay minerals tend to remain in suspension, since like charges repel. These forces of repulsion are strong relative to the van der Waals' attractive forces (see Box 3.10) and prevent particles from aggregating and sinking. It follows that anything which neutralises surface charges will allow particles to flocculate. Many colloids flocculate in an electrolyte, and seawater — a much stronger electrolyte than riverwater — fulfils this role in estuaries. The cations in seawater are attracted to the negative charges on clay surfaces. The cations form a mobile layer in solution adjacent to the clay surface (Fig. 4.1) and the combined 'electrical double layer' is close to being electrically neutral. Adjacent particles can then approach each other and aggregate. In nature, this simple

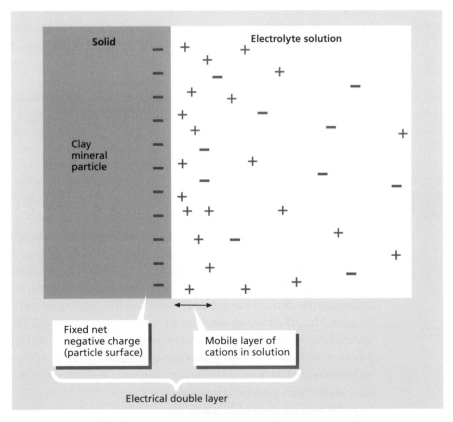

Fig. 4.1 Electrical double layer, comprising a fixed layer of negative charge on the particle and a mobile ionic layer in solution. The latter is caused because positive ions are attracted to the particle surface. Note that with increasing distance from the particle surface the solution approaches electrical neutrality. After Raiswell *et al.* (1980).

explanation is vastly complicated by the presence of organic and oxyhydroxide coatings on particles.

Sedimentation in estuaries is localised to the low-salinity region by the physical and chemical effects discussed above. The sediment is, however, continuously resuspended by tidal currents, moving upstream on incoming tides and downstream on the ebb. The net effect is to produce regions of high concentrations of suspended particulate matter, known as turbidity maxima. The turbidity maximum is an important region because many reactions in environmental chemistry involve exchange of species between dissolved and particulate phases. These reactions can significantly affect fluxes of riverine material to the oceans and therefore must be quantified in order to understand global element cycling.

4.2.2   Mixing processes in estuaries

Water flow in estuaries is not unidirectional; it is subject to the reversing flows of tides. There is therefore no constant relationship between a fixed geographic point and water properties (e.g. calcium ion (Ca^{2+}) concentration). For this reason, data collected in estuaries are usually compared with salinity (Box 4.1) rather than location. The underlying assumption is that salinity in an estuary is simply the result of physical mixing and not of chemical changes. If the estuary has just one river entering it and no other inputs, the behaviour of any component can be assessed by plotting its concentration against salinity.

If the concentration of the measured component is, like salinity, controlled by simple physical mixing, the relationship will be linear (Fig. 4.2). This is called conservative behaviour and may occur with riverine concentrations higher than, or

---

Box 4.1

Salinity

Salinity is defined as the weight in grams of inorganic ions dissolved in 1 kg of water. Seven ions constitute more than 99% of the ions in seawater and the ratios of these ions are constant throughout the world oceans. Consequently, the analysis of one ion can, by proportion, give the concentration of all the others and the salinity. The density of seawater and light and sound transmission all vary with salinity.

Salinity is measured by the conductance of electrical currents through the water (conductivity). Measured values are reported relative to that of a known standard; thus salinity has no units — although, in many older texts, salinities are reported in units of parts per thousand (ppt or ‰) or grams per litre.

Open-ocean waters have a narrow range of salinities (32–37) and most are near 35. In estuaries, values fall to less than 1 approaching the freshwater endmember. In hypersaline environments salinities can exceed those of seawater, reaching values greater than 300.

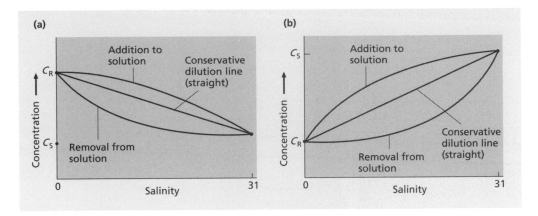

Fig. 4.2 Idealised plots of estuarine mixing illustrating conservative and non-conservative mixing. C_R and C_S are the concentrations of the ions in river and seawater respectively. After Burton & Liss (1976).

lower than, in seawater (Fig. 4.2). In contrast, if there is addition of the component, unrelated to salinity change, the data will plot above the conservative mixing line (Fig. 4.2). Similarly, if there is removal of the component, the data will plot below the conservative mixing line (Fig. 4.2). In most cases, removal or input of a component will occur at low salinities and the data will approach the conservative line at higher salinity (Fig. 4.3). Extrapolation of such a 'quasi-conservative line' back to zero salinity can provide, by comparison with the measured zero salinity concentration, an estimate of the extent of removal or release of the component (Fig. 4.3).

4.2.3   Halmyrolysis and ion exchange in estuaries

The electrochemical reactions that impinge on river-borne clay minerals carried into seawater do not finish with flocculation of particles and sedimentation of aggregates. The capacity for ion exchange in clay minerals means that their transport from low-ionic-strength, Ca^{2+} and HCO_3^- dominated riverwater, to high-ionic-strength, sodium chloride (NaCl)-dominated seawater demands reaction with the new solution to regain chemical equilibrium (see Box 2.4). The process by which terrestrial materials adjust to marine conditions has been called 'halmyrolysis', derived from Greek roots *hali* (sea) and *myros* (unguent), literally 'to anoint with the sea'. Halmyrolysis is imprecisely defined but we will consider it to encompass all those reactions that affect a particle in seawater before burial in sediment.

Various measurements of cation exchange on river clays in seawater have shown that clay minerals exchange adsorbed Ca^{2+} for Na^+, potassium ions (K^+) and magnesium ions (Mg^{2+}) from seawater (see Section 4.4.3). In general, components with a high affinity for solid phases, such as dissolved phosphorus (P) or iron (Fe) (Fig. 4.3), are removed from solution. Thus the rules of ionic behaviour arising from consideration of charge/ionic radius (z/r) ratios (see Section 3.7.1) are helpful in understanding chemical behaviour in estuarine environments, as well as in weathering.

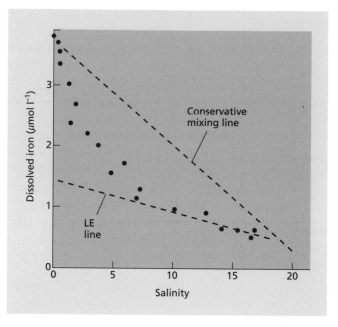

Fig. 4.3 Dissolved iron versus salinity in the Merrimack Estuary (eastern USA), illustrating non-conservative behaviour. Linear extrapolation (LE) of high-salinity iron data to zero salinity gives an estimate of 60% low-salinity removal of iron. After *Geochimica Cosmochimica Acta*, **38**, Boyle, E.A., Collier, R., Dengler, A.T., Edmond, J.M., Ng, A.C. & Stallard, R.F. On the chemical mass-balance in estuaries, 1719–1728, Copyright 1974, with kind permission from Elsevier Science Ltd, The Boulevard, Langford Lane, Kidlington OX5 1GB, UK.

4.2.4 Microbiological activity in estuaries

As in most environments, biological, particularly microbial, processes are important in estuaries. In many estuaries the high particulate concentrations make waters too turbid to allow phytoplankton growth. However, in shallow or low-turbidity estuaries, or at the seaward end of estuaries where suspended solid concentrations are low, sunlight levels may be sufficient to sustain phytoplankton growth. Estuaries frequently provide safe, sheltered harbours, often centres of trade and commerce. As a result, in developed and developing countries estuary coasts are often sites of large cities. Discharge of waste, particularly sewage, from the population of these cities increases nutrient concentrations and, where light is available, large amounts of primary production occur (see Section 3.7.5). In the dynamic environment of an estuary, dilution of phytoplankton-rich estuarine water with offshore low-phytoplankton waters, occurs at a faster rate than cells can grow (phytoplankton populations under optimum conditions can double on timescales of a day or so). Thus, phytoplankton populations are often limited by this dilution process, rather than by nutrient or light availability.

The extent of nutrient removal that can occur in estuaries is illustrated for silicate (SiO_2) in the estuary of the River Great Ouse in eastern England (Fig. 4.4). In this example there are a number of parameters whose values point to the role of phytoplankton in nutrient removal. Most importantly, silicate removal is related to high particulate chlorophyll concentrations, oxygen (O) supersaturation (arising from photosynthesis — see eq. 3.28) and removal of other nutrients. During winter, silicate removal ceases when chlorophyll levels are low, allowing oxygen concentrations to fall to lower levels.

In Chesapeake Bay, a large estuary on the east coast of the USA, phytoplankton blooms in the high-salinity part of the estuary generate large amounts of organic matter which sink into the deep waters. The deep waters are isolated from surface waters by thermal stratification and the breakdown of the phytoplankton debris results in occasional, seasonal, dissolved oxygen depletion and consequent death of fish and invertebrates.

The processes that concentrate sediments in estuaries also concentrate particulate organic matter. If large amounts of organic matter are present in an estuary, oxygen consumption rates, resulting from aerobic bacterial consumption of organic matter, can exceed the rate at which oxygen is supplied. This results in decreasing dissolved oxygen concentrations.

The discharge of sewage from cities often causes low or zero dissolved oxygen concentrations. A particularly well-documented example is the River Thames in southern England. Throughout the eighteenth century, the wastes of the population of London were dumped in streets and local streams. During the early nineteenth century, public health improvements led to the development of sewers and the discharge of sewage directly into the Thames. The result was an improvement of local sanitation but massive pollution of the Thames. Salmon and almost all other animals disappeared and the river had to be abandoned as a water supply. The literature of the time refers to the foul smells, and public concern prompted the

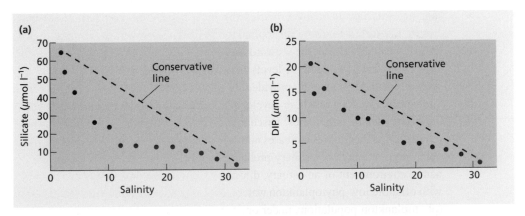

Fig. 4.4 (a) Dissolved silicate and (b) dissolved inorganic phosphorus plotted against salinity in the Great Ouse Estuary (eastern England), illustrating non-conservative removal.

development of sewage treatment works and also routine monitoring of environmental conditions in the estuary. The sewage treatment allowed dissolved oxygen concentrations to rise and fish returned to the estuary. However, as the population of London grew, the treatment system became overloaded and environmental conditions in the estuary deteriorated once more. The decrease of dissolved oxygen concentrations and their subsequent increase, arising from improved sewage treatment in the 1950s, are illustrated in Fig. 4.5. The story of the Thames indicates the interaction between public health improvements and pollution; it also shows that some environmental problems are at least in part reversible, given political will and economic resources.

4.3 Major ion chemistry of seawater

Having examined the chemistry of estuarine environments, we now turn to global chemical cycling in the open ocean. This chapter began by noting that the major ion chemistry of seawater is different from that of continental surface waters (Table 4.1). There are three principal features which mark clearly this difference.
1 The high ionic strength of seawater (see Fig. 3.24), containing about 35 g l^{-1} of salts (Box 4.1).
2 The chemical composition of seawater, with highly abundant Na$^+$ and Cl$^-$ (Table 4.1).
3 Seawater has remarkably constant relative concentrations of the major ions in all of the world's oceans.

The latter feature is well illustrated by data which show the K$^+$: Cl$^-$ ratio measured in oceanic areas from the Arctic Ocean to the Black Sea to be almost invariant within analytical errors at 0.0205 ± 0.0007.

In the oceans, bicarbonate ions (HCO$_3^-$) and Ca^{2+} are biologically cycled, causing vertical gradients in their ratios relative to the other major ions. However, the differences in the ratios are small — less than 1% for calcium. There is also evi-

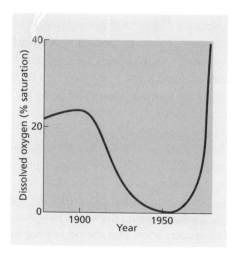

Fig. 4.5 Average autumn dissolved oxygen concentrations in the tidal River Thames. After Wood (1982).

Table 4.1 Major ion composition of freshwater and seawater in mmol l^{-1}. Global average riverwater data from Berner & Berner (1987), seawater data from Broecker & Peng (1982)

	Riverwater	Seawater
Na^+	0.23	470
Mg^{2+}	0.14	53
K^+	0.03	10
Ca^{2+}	0.33	10
HCO_3^-	0.85	2
SO_4^{2-}	0.09	28
Cl^-	0.16	550
Si	0.16	0.1

dence that the major ion composition of seawater has varied only modestly over many millions of years (Box 4.2), suggesting that controls on ocean composition are linked to very long-term geochemical cycling.

4.4   Chemical cycling of major ions

Residence times of the major ions in seawater (Box 4.3) are important indicators of the way chemical cycling operates in the oceans. These residence times are all very long (10^4 to 10^8 years), similar to or longer than the water itself (around 3.8×10^4 years). Long residence times mean there is ample opportunity for ocean currents to mix the water and constituent ions thoroughly. This ensures that changes in ion ratios arising from localised input or removal processes are smoothed out. It is the long residence times of the ions that create the very constant ion ratios in seawater. The residence times result from the high solubility of the ions and hence their z/r ratios (see Section 3.7.1). Other cations with similar z/r ratios will also have long oceanic residence times (e.g. caesium ion (Cs^+)), but these are not major ions in seawater because of their low crustal abundances. Chloride is an interesting exception as it is abundant in seawater, has a long residence time and yet has a low crustal abundance. Most of this Cl^- was degassed from the Earth's mantle as hydrogen chloride (HCl) very early in Earth history (see Section 1.3.1) and has been recycled in a hydrosphere–evaporite cycle since then (see Section 4.4.2).

The residence times in Box 4.3 are based on riverwater being the only input of ions to the oceans. This is a simplification as there are also inputs from the atmosphere and from hydrothermal (hot water) processes at mid-ocean ridges (Fig. 4.6). For major ions, rivers are the main input, so the simplification in Box 4.3 is valid. For trace metals, however, atmospheric and mid-ocean ridge inputs are important and cannot be ignored in budget calculations (see Section 4.5).

The long residence times of the major ions compared with the water mean that seawater is a more concentrated solution than riverwater. However, the different ionic ratios of seawater and riverwater show that the oceans are not simply the

Box 4.2

Constancy of major ion chemistry of seawater on geological timescales

The evidence that the salinity and ionic composition of seawater have remained reasonably constant over at least the last 900 million years comes from ancient marine evaporite deposits. Evaporites are salts which have precipitated naturally from evaporating seawater in basins largely cut off from the open ocean.

Over the last 900 million years, marine evaporites have normally begun with a gypsum–anhydrite section ($CaSO_4.2H_2O$–$CaSO_4$), followed by a halite (NaCl) sequence. Bittern salts (named due to their bitter taste) have precipitated from the final stages of evaporation and have variable composition, including magnesium salts, bromides, potassium chloride (KCl) and more complex salts, depending on conditions of evaporation (Fig. 1).

The order of precipitates is the same as that seen in modern marine evaporites and can be reproduced by experimental evaporation of seawater. This sequence of salt precipitation sets limits on the possible changes of major ion compositions in seawater, since changes beyond these limits would have resulted in different sequences of precipitation.

Calculations demonstrating the actual limits on changes in the major ion chemistry of seawater imposed by the evaporite precipitation sequence are beyond the scope of this book. However, a number of simple observations give some idea of possible variations. For example, doubling calcium ion (Ca^{2+}) seawater

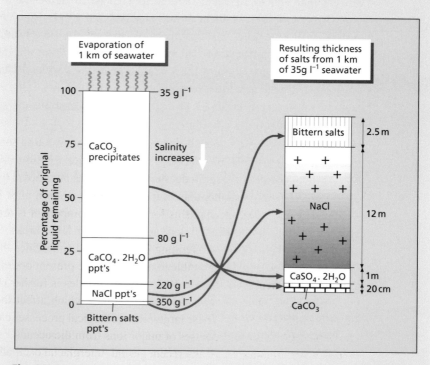

Fig. 1 Successional sequence and approximate thickness of salts precipitated during the evaporation of 1 km of seawater. After Scoffin (1987).

**Box 4.2
Cont.**

concentrations at present sulphate ion (SO_4^{2-}) concentrations would not affect the sequence, whereas tripling the Ca^{2+} concentration would. Similarly, halving or doubling present-day potassium ion (K^+) concentrations would result in the formation of some very unusual bittern salts, not seen in the geological record.

Ideas about variations in sodium ion (Na^+) and chloride ion (Cl^-) concentrations are based on ancient halite inventories. The total volume of known halite deposits amounts to about 30% of the NaCl content of the present oceans. If all of this salt were added to the present oceans, the salinity of seawater would increase by about 30%, setting an upper limit. However, the ages of major halite deposits are reasonably well dispersed through geological time, suggesting that there was never a time when all of these ions were dissolved in seawater.

Setting lower limits on Na^+ and Cl^- concentrations in seawater can be estimated by considering the larger evaporite deposits in the geological record. For example, in Miocene times (5–6 million years ago) about 28×10^{18} mol of NaCl was deposited in the Mediterranean–Red Sea basins. This volume of salt represents just 4% of the present mass of oceanic NaCl. This suggests that periodic evaporite-forming events are only able to decrease Na^+ and Cl^- concentrations of seawater by small amounts. It has been suggested that the salinity of seawater has declined in 'spurts' from 45 to 35 g l^{-1} over the last 570 million years. During this time the formation of Permian-aged salts alone (280–230 million years ago) may have caused a 10% decrease in salinity, possibly contributing to the extinction of many marine organisms at the end of this period.

Overall, these types of constraints suggest that the major ion chemistry of seawater has varied only modestly (probably by no more than a factor of 2 for each individual ion) during the last 900 million years, or so (slightly less than a quarter of geological time).

result of riverwater filling the ocean basins, even if the resulting solution has been concentrated by evaporation. Although major ion residence times are all long, they vary over four orders of magnitude, showing that rates of removal for specific ions are different. Processes other than evaporative concentration must be operating.

Identifying removal mechanisms for a specific component is difficult because removal processes are usually slow and occur over large areas. Some removal processes are very slow — operating on geological timescales of thousands or millions of years — and impossible to measure in the present oceans. The requirement to study element cycles on geological timescales is further complicated by processes like climate change and plate tectonics which affect the geometry of ocean basins and sealevel. These large-scale geological processes can have significant effects on removal processes of major ions from the oceans.

The effects of the geologically recent glacial–interglacial oscillations during the Quaternary period (the last 2 million years) are particularly relevant. Firstly, the rapid rise in sealevel over the last 11 000 years, following the melting of polar ice

Box 4.3

Residence times of major ions in seawater

The total volume of the oceans is 1.37×10^{21} l and the annual river flow to the oceans is 3.6×10^{16} l year^{-1}. The residence time of water in the oceans is therefore

$$\frac{\text{Inventory}}{\text{Input}} = \frac{1.37 \times 10^{21}}{3.6 \times 10^{16}} = 3.8 \times 10^4 \text{ years} \qquad \text{eq. 1}$$

Applying this approach to the data in Table 4.1, it is straightforward to calculate the residence times of the major ions, assuming that:
1 dissolved salts in rivers are the dominant sources of major ions in seawater;
2 steady-state conditions apply (see Section 2.3).
The first assumption is probably valid, since the other sources listed in Table 4.2 do not greatly alter the results derived by considering rivers alone. The issue of steady state cannot be verified for very long (millions of years) timescales, but the geological evidence does suggest that the concentration of major ions in seawater has remained broadly constant over very long time periods (Box 4.2). As an example of the residence time calculation, consider sodium (Na^+).

$$\begin{aligned}
\text{Input} &= \text{water flow in rivers} \times \text{river concentration} \\
&= 3.6 \times 10^{16} \times 0.23 \times 10^{-3} \text{ mol year}^{-1} \\
&= 8.28 \times 10^{12} \text{ mol year}^{-1} \qquad \text{eq. 2}
\end{aligned}$$

$$\begin{aligned}
\text{Inventory} &= \text{water content of the oceans} \times \text{ocean concentration} \\
&= (1.37 \times 10^{21}) \times (470 \times 10^{-3}) \text{ mol} \\
&= 644 \times 10^{18} \text{ mol} \qquad \text{eq. 3}
\end{aligned}$$

$$\text{Residence time} = \frac{644 \times 10^{18} \text{ mol}}{8.28 \times 10^{12} \text{ mol year}^{-1}} = 78 \times 10^6 \text{ years} \qquad \text{eq. 4}$$

Table 1 Residence times of major ions in seawater

Ion	Residence time (10^6 years)
Na^+	78
Mg^{2+}	14
K^+	13
Ca^{2+}	1.1
HCO_3^-	0.09
SO_4^{2-}	12
Cl^-	131

accumulated during the last glacial period, has flooded former land areas to create large, shallow, continental shelves, areas of high biological activity and accumulation of biological sediments (see Section 4.2.4). Also, the unconsolidated glacial

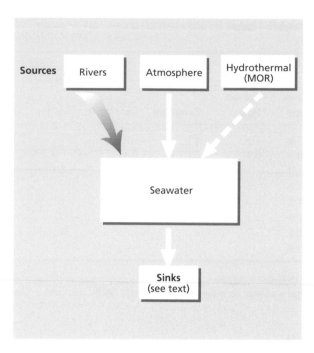

Fig. 4.6 Simple box model summarising material inputs to seawater. Size of arrow indicates relative importance of input.

sediments which mantle large areas of northern hemisphere (temperate-arctic zone) land surfaces are easily eroded. This results in high particulate concentrations in rivers, which carry material to estuaries and continental shelves. This enhanced sediment supply results in correspondingly high detrital sediment–seawater inter-actions, increasing the importance of removal processes such as ion exchange.

Despite these complications, the main removal mechanisms of major ions from seawater are known (Table 4.2). Quantifying the importance of each mechanism is less easy and the uncertainty of data in Table 4.2 should not be forgotten. In the following section we outline the important removal processes for major ions in seawater.

4.4.1   Sea-to-air fluxes

Sea-to-air fluxes of major ions are caused by bubble bursting and breaking waves at the sea surface. These processes eject sea salts into the atmosphere, the majority of which immediately fall back into the sea. Some of these salts are, however, transported over long distances in the atmosphere and contribute to the salts in riverwater (see Section 3.7.2). These airborne sea salts are believed to have the same relative ionic composition as seawater. In terms of global budgets, airborne sea salts are an important sink only of Na^+ and Cl^- from seawater.

Table 4.2 Simplified budget for major ions in seawater. All values are in 10^{12} mol year $^{-1}$. After Berner & Berner (1987)

Ion	River input	Sea–air fluxes	Removal/source*					
			Evaporites	CE–clay	$CaCO_3$	Opaline silica	Sulphides	MOR
Cl^-	5.8	1.1	4.7					
Na^+	8.3	0.9	4.7	0.8				1.6
Mg^{2+}	5.0			0.1	0.6			4.9
SO_4^{2-}	3.2		1.2				1.2	
K^+	1.1			0.1				–0.8
Ca^{2+}	11.9		1.2	–0.5	17			–4.8
HCO_3^-	30.6				34		–2.4	
Si	5.8					7.0		–1.1

* Minus sign indicates a source.

CE–clay, cation exchange on estuarine clay minerals; MOR, mid-ocean ridge and other seawater–basalt interactions.

4.4.2   Evaporites

Evaporation of seawater will precipitate the constituent salts, the so-called evaporite minerals, in a predictable sequence (see Box 4.2). This sequence starts with the least soluble salts and finishes with the most soluble (see Box 3.8). If approximately half (47%) of the water volume is evaporated, $CaCO_3$ precipitates (see eq. 4.3). With continued evaporation and approximately fourfold increase in salinity, $CaSO_4.2H_2O$ (gypsum) precipitates.

$$Ca^{2+}_{(aq)} + SO^{2-}_{4(aq)} + 2H_2O_{(1)} \rightleftharpoons CaSO_4.2H_2O_{(s)} \qquad \text{eq. 4.1}$$

Once 90% of the water (H_2O) has been evaporated, at dissolved salt concentrations around 220 g l^{-1}, NaCl precipitates,

$$Na^+_{(aq)} + Cl^-_{(aq)} \rightleftharpoons NaCl_{(s)} \qquad \text{eq. 4.2}$$

and, in addition, some magnesium (Mg) salts begin to crystallise; if evaporation continues, highly soluble potassium (K) salts precipitate (see Box 4.2).

The problem with invoking evaporation as a removal mechanism for ions in seawater is that there are currently very few environments in which evaporite salts are accumulating to a significant extent. This is because enormous volumes of seawater need to be evaporated before the salts become concentrated enough to precipitate. Clearly this cannot occur in the well-mixed open oceans, where net evaporative water loss is roughly balanced by resupply from continental surface waters (river flux). This implies that evaporative concentration of seawater can only occur in arid climatic regions within basins largely isolated from the open ocean and other sources of water supply. There are no modern examples of such basins; modern evaporite deposits are two to three orders of magnitude smaller than ancient deposits and are restricted to arid tidal flats and associated small salt ponds — for example, on the Trucial Coast of the Arabian Gulf. Note, however, that large evaporite deposits do exist in the geological record, the most recent example resulting from the drying out of the Mediterranean Sea in late Miocene times (about 5–6 million years ago).

As Cl^- has a very long oceanic residence time, the sporadic distribution of evaporite-forming episodes (see Box 4.2), integrated over million-year timescales, results in only quite small fluctuations in the salinity of seawater. However, the lack of major evaporite-forming environments today suggests that both Cl^- and sulphate (SO_4^{2-}) are gradually accumulating in the oceans until the next episode of removal by evaporite formation.

In Table 4.2 the amount of Cl^- removed from seawater by evaporation has been set to balance the input estimate (there being no other obvious Cl^- sink except sea-to-air fluxes and burial of pore water (see Section 4.4.8); this in turn dictates that the same amount of Na^+ is removed to match the equal ratios of these ions in NaCl. The figure for SO_4^{2-} removal by evaporation (Table 4.2) is plausible, albeit poorly constrained. Again, the SO_4^{2-} estimate dictates an equal removal of Ca^{2+} ions to form $CaSO_4.2H_2O$.

4.4.3 Cation exchange

Ion exchange processes on clay minerals moving from riverwater to seawater (see Section 4.2.3) remove about 26% of the river flux of Na^+ to the oceans and are significant removal processes for K^+ and Mg^{2+} (Table 4.3). Clay minerals are also a significant source of Ca^{2+} to the oceans, adding an extra 8% to the river flux (Table 4.3). These modern values are, however, thought to be double the long-term values due to the effects of abnormally high postglacial suspended-solid input rates (see also Section 4.4.4). The modern values are halved in Table 4.2 to account for this.

4.4.4 Carbonate precipitation

It is difficult to calculate accurately whether seawater is supersaturated or undersaturated with respect to $CaCO_3$. There are a number of different approaches to the problem, but all are based on the following equilibrium relationship, which describes the precipitation (forward reaction) or dissolution (back reaction) of $CaCO_3$.

$$Ca^{2+}_{(aq)} + 2HCO^-_{3(aq)} \rightleftharpoons CaCO_{3(s)} + CO_{2(g)} + H_2O_{(l)} \qquad \text{eq. 4.3}$$

Seawater is a concentrated and complex solution in which the ions are close together, compared with those in a more dilute solution. Electrostatic interaction occurs between closely neighbouring ions and this renders some of these ions 'inactive'. We are interested in the available or 'active' ions (Box 4.4) and we correct for this effect, using activity coefficients (Box 4.4).

Table 4.3 Additions to the river flux from ion exchange between river-borne clay and seawater. Drever, J.I, Li, Y.-H. & Maynard, J.B. in *Chemical Cycles and the Evolution of the Earth*, ed. by Gregor, C.B., Garrels R.M., Mackenzie, F.T. & Maynard, J.B. pp. 17–53, ©1988. Reprinted by permission of John Wiley & Sons, Inc.

	Laboratory studies (10^{12} mol year^{-1})	Amazon (10^{12} mol year^{-1})	Average* (10^{12} mol year^{-1})	Percentage of river flux†
Na^+	−1.58	−1.47	−1.53 ± 0.06	26
K^+	−0.12	−0.27	−0.20 ± 0.08	17
Mg^{2+}	−0.14	−0.49	−0.31 ± 0.08	7
Ca^{2+}	0.86	1.05	0.96 ± 0.10	8

* The averages are based on modern suspended sediment input to the oceans, which is probably double the long-term input rate (see Section 4.4). These values are thus halved when used in Table 4.2.
† Corrected for cyclic salts and pollution.
Minus sign indicates removal from seawater.

Activity coefficients are notoriously difficult to measure in complex solutions like seawater, but are thought to be around 0.26 for Ca^{2+} and around 0.20 for carbonate (CO_3^{2-}). Measured Ca^{2+} and CO_3^{2-} ocean surface concentrations are 0.01 and 0.000 29 mol l^{-1} respectively and thus the ion activity product (IAP) (Box 4.4) can be calculated.

$$IAP = aCa^{2+} \times aCO_3^{2-}$$
$$= 0.01 \times 0.26 \times 0.000\ 29 \times 0.2 \qquad \text{eq. 4.4}$$
$$= 1.5 \times 10^{-7}\,mol^2\,l^{-2}$$

This value is much greater than the solubility product (see Box 3.8) of calcite ($CaCO_3$), which is 4.5×10^{-9} mol^2 l^{-2}. The degree of saturation (see Box 3.8) is:

$$\text{Degree of saturation } \Omega = \frac{IAP}{K_{sp}} \qquad \text{eq. 4.5}$$

An Ω value of 1 indicates saturation, values > 1 indicate supersaturation, and values < 1 indicate undersaturation. Using the values for calcite above we get:

$$\Omega = \frac{1.5 \times 10^{-7}}{4.5 \times 10^{-9}} = 33.3 \qquad \text{eq. 4.6}$$

implying that seawater is supersaturated with respect to calcite. This approach neglects the effects of ion pairing (Box 4.5). Allowance for ion pairing indicates that about 90% of the calcium in seawater is present as the free ion with the remainder present as $CaSO_4^0$ and $CaHCO_3^+$ ion pairs. For CO_3^{2-} it is calculated that only about 10% exists as the free ion with the remainder in ion pairs with Mg^{2+}, Ca^{2+} and Na^+. Applying these corrections decreases the IAP in the following way.

$$IAP = 1.5 \times 10^{-7} \times 0.9 \times 0.1$$
$$= 1.35 \times 10^{-8}\,mol^2\,l^{-2} \qquad \text{eq. 4.7}$$

Box 4.4

Active concentrations

The active concentration or *activity* of an ion becomes an important consideration in concentrated and complex solutions like seawater. Ions in a concentrated solution are sufficiently close to one another for electrostatic interactions to occur. These interactions reduce the effective concentration of ions available to participate in reactions. In order to accurately predict chemical reactions in a concentrated solution, we need to account for the reduction in effective concentration, which is done using an activity coefficient (γ).

activity = concentration $\times \gamma$ eq. 1

Equation 1 shows that units of activity and concentration are proportional; in other words γ can be regarded as a proportionality constant. These constants, which vary between 0 and 1, can be calculated experimentally or theoretically and are quite well known for some natural solutions. Having said this, measuring γ in complex solutions like seawater has proved very difficult. Most importantly for our purposes, as solution strength approaches zero, γ approaches 1. In other words, in very dilute solutions (e.g. rainwater), activity and concentration are effectively the same.

In this book activity is expressed in units of mol l^{-1} in the same way as concentration. Activity is denoted by a, and the product of multiplying ion activities is known as the ion activity product (IAP), as in eq. 4.4.

We should also note that thermodynamic equilibrium constants are measured in terms of activity. Thus, measured concentrations of any chemical species should be converted to activities before comparison with thermodynamic data.

Now the calculation of saturation state becomes:

$$\Omega = \frac{1.35 \times 10^{-8}}{4.5 \times 10^{-9}} = 3$$ eq. 4.8

On this basis, surface seawater is about three times supersaturated with respect to calcite. Various approaches to this problem, all with inherent assumptions, yield Ω values between three and six times supersaturated. Overall, we can conclude that surface seawater is supersaturated with respect to calcite and we might expect $CaCO_3$ to precipitate spontaneously.

The evidence from field studies is somewhat contrary to the predictions based on equilibrium chemistry. Abiological precipitation of $CaCO_3$ seems to be limited, restricted to geographically and geochemically unusual conditions. The reasons why carbonate minerals are reluctant to precipitate from surface seawater are still poorly understood, but probably include inhibiting effects of other dissolved ions and compounds. Even where abiological precipitation is suspected — for example, the famous ooid shoals and whitings of the Bahamas (Box 4.6) — it is often difficult to discount the effects of microbial mediation in the precipitation process.

Box 4.5

Ion interactions and ion pairing

When ionic salts dissolve in water, the salts dissociate to release the individual ions. The charged ions attract the polar water molecules such that a positively charged ion will be surrounded most closely by oxygen atoms of the water molecules (Fig. 1). Thus, ions are not free in solution, but interacting with water molecules. For example, the hydrogen ion (H^+) is hydrated to form H_3O^+. For simplicity in chemical equations the simple H^+ notation is used.

In addition to interaction with water molecules, individual ions may interact with other ions to form *ion pairs*. For example, dissolved sodium and sulphate ions can interact to form a sodium sulphate ion pair.

$$Na^+_{(aq)} + SO^{2-}_{4(aq)} \rightleftharpoons NaSO^-_{4(aq)}$$ eq. 1

The extent of this interaction varies for different ions and is measured by an equilibrium constant. In this case:

$$K = \frac{aNaSO^-_4}{aNa^+.aSO^{2-}_4}$$ eq. 2

If all the relevant equilibrium constants are known, together with the amounts of the ions present, the proportions of the various ions associated with each ion pair can be calculated. The results of such an analysis for seawater (Table 1) show that ion pairing is common. A full analysis of the properties of seawater requires that these species are taken into account.

Since anions are present at lower concentrations than cations (except for chloride (Cl^-)), ion pairing has a proportionately greater effect on anions relative to cations. The extent of ion pairing is dependent on temperature, pressure and salinity.

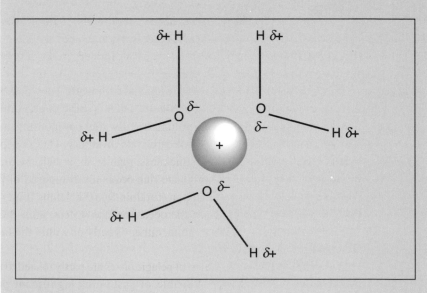

Fig. 1. Ionic salts dissolved in water.

Box 4.5 Cont.

Table 1 Percentage of each major ion in seawater present in various ion pairs

	Na$^+$	Mg^{2+}	Ca^{2+}	K$^+$
Free ion	98	89	89	99
MSO$_4$	2	10	10	1
MHCO$_3$	—	1	1	—
MCO$_3$	—	—	—	—

	Cl$^-$	SO$_4^{2-}$	HCO$_3^-$	CO$_3^{2-}$
Free ion	100	39	81	8
NaX	—	37	11	16
MgX	—	20	7	44
CaX	—	4	4	21
KX	—	1	—	—
Mg$_2$CO$_3$	—	—	—	7
MgCaCO$_3$	—	—	—	4

M, cation; X, anion.
Blanks refer to species present at less than 1%.

Volumetrically, the biological removal of Ca^{2+} and HCO$_3^-$ ions, synthesised into the skeletons of organisms, is much more important. It is tempting to assume that obvious features, such as the substantial coral reefs of tropical and subtropical oceans (e.g. the Australian Great Barrier Reef) and other shallow marine skeletal carbonate, represent major sites of removal. Indeed, in the modern oceans, the large continental shelf areas created by sealevel rise in the last 11 000 years probably do account for about 45% of global carbonate deposition. However, over the last 150 million years it can be shown that it is carbonate sedimentation in the deep oceans which has been volumetrically more important, accounting for between 65 and 80% of the global carbonate inventory. These deep-sea deposits, which average about 0.5 km in thickness, mantle about half the area of the deep ocean floor (Fig. 4.7). The carbonate rich oozes are composed of phytoplankton (coccolithophores) and zooplankton (foraminifera) skeletons (Fig. 4.8). Although these pelagic organisms live in the ocean surface waters, after death their skeletons sink through the water column, either directly or within the faecal pellets of zooplankton.

The controls on the distribution of pelagic oozes are partly related to the availability of nutrients, which must be capable of sustaining significant populations of phytoplankton (see Section 3.7.4). More important, however, is the dissolution of CaCO$_3$ as particles sink into ocean deep waters. In the deep ocean, carbon dioxide (CO$_2$) con-

Box 4.6

Abiological precipitation of calcium carbonate

Where a skeletal source cannot be identified, calcium carbonate ($CaCO_3$) grains and fine-grained muds may be of abiological origin. The most famous occurrences occur in shallow, warm, saline waters of the Bahamas and the Persian Gulf. In these areas two distinctive morphologies are present, ooids and needle muds (Fig. 1).

Ooids are formed by aggregation of aragonite* crystals around a nucleus, usually a shell fragment or pellet. Successive layers of aragonite precipitation build

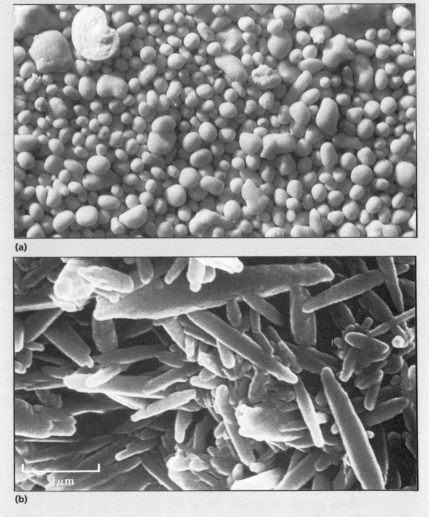

(a)

(b)

Fig. 1 (a) Ooid-rich sediment from the Great Bahama Bank. Individual grains are typically 1 mm in diameter. Photograph courtesy of J. Andrews. (b) Scanning electron microscope photograph of aragonitic needles from the Great Bahama Bank. Scale bar = 1 μm. Photograph courtesy of I.G. Macintyre & Reid R.P. (1992) *Journal of Sedimentary Petrology* **62**, 1095–1097, Society for Sedimentary Geology, Tulsa.

Box 4.6
Cont.

up a concentric structure, which may vary in size from about 0.2 to 2.0 mm in diameter. Needle muds are also aragonitic; typically each needle is a few micrometres wide and tens of micrometres long.

It has long been thought that the warm, shallow, saline waters where these deposits are found favour increased concentrations of carbonate ions (CO_3^{2-}), increasing the ion activity product of $aCa^{2+}.aCO_3^{2-}$ such that precipitation of $CaCO_3$ occurs. The formation of ooids probably requires fairly agitated, wind- or wave-stirred waters, allowing periodic suspension of the grains into the CO_3^{2-}-saturated water, whereas aragonitic needles may precipitate as clouds of suspended particles, known as whitings.

There has been, and still is, much debate about the origins of these particles. Firstly, it is difficult to disprove the effects of microbial mediation in their formation. Thus we might regard the grains as non-skeletal, while accepting a possible microbial influence. Secondly, various geochemical and mineralogical studies have produced equivocal results in attempting to demonstrate an abiological origin. Having said this, recent work based on crystal morphology and strontium substitution lends support for inorganic precipitation of needle muds.

Despite the considerable interest these phenomena provoke, we should remember that they are of minor significance to the modern oceanic CO_3^{2-} budget. The relative importance of inorganic CO_3^{2-} in the geological past is more difficult to assess, but may have been more significant before the evolution of shelly organisms about 570 million years ago.

* Aragonite and calcite are known as polymorphs of $CaCO_3$. Both minerals have the formula $CaCO_3$, but they differ slightly in the structural arrangement of atoms.

centrations increase, particularly in the deep Pacific, as a result of the decomposition of sedimenting organic matter. Decreased temperature and increased pressure also promote dissolution of $CaCO_3$ favouring the reverse reaction in eq. 4.3.

By mapping the depth at which carbonate sediments exist on the floors of the oceans, it has been possible to identify the level where the rate of supply of biogenic CO_3^{2-} is balanced by the rate of solution. This depth, known as the calcite compensation depth (CCD), is variable in the world's oceans, depending on the degree of CO_3^{2-} undersaturation in the deep waters (Fig. 4.9). In the Atlantic Ocean the CCD is at about 4.5 km depth; above the CCD, at about 4 km depth in the Atlantic, there is a critical depth, known as the lysocline (Fig. 4.9). Here the rate of calcite dissolution increases markedly and all but the most robust particles dissolve rapidly. It is estimated that about 80% of the $CaCO_3$ settling into deep waters is dissolved, either during transit through the water column or on the seabed. As a consequence, pelagic carbonate deposits are most common on the shallower parts of the deep ocean floors (Fig. 4.9) or on topographic highs which project above the CCD.

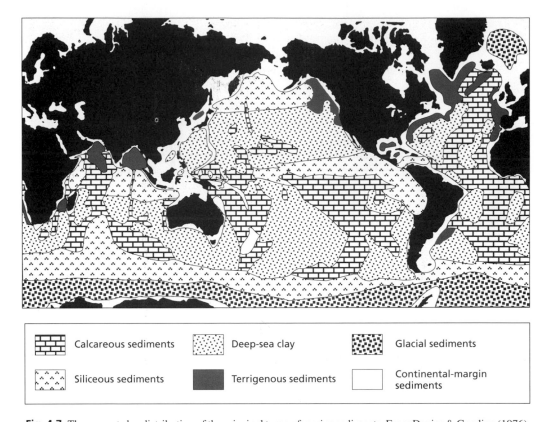

Fig. 4.7 The present-day distribution of the principal types of marine sediments. From Davies & Gorsline (1976).

Planktonic coccolithophores and foraminifera did not evolve until the mid-Mesozoic (about 150 million years ago), whereas shallow-water shelly organisms are known to have existed throughout Phanerozoic time (570 million years to the present day). This means that the locus of carbonate deposition shifted to the deep oceans only in the last quarter of Phanerozoic time.

The removal of Ca^{2+} by $CaCO_3$ precipitation can be estimated directly from the abundance of carbonate-rich ocean sediments and their sedimentation rates (see Table 4.2). From eq. 4.3, we see that two moles of HCO_3^- are removed with each mole of Ca^{2+}, a process that releases dissolved CO_2 into seawater, ultimately to be returned to the atmosphere. Carbonates also incorporate a small, but significant amount of Mg^{2+} by isomorphous substitution for Ca^{2+} (see Box 3.9) and this is used to derive the Mg^{2+} removal in Table 4.2.

Finally, we should note that seawater is roughly in equilibrium with calcite and that the presence of carbonate sediments in rocks of all ages suggests that it has been throughout much of Earth history. Oceanic pH is unlikely to have fallen

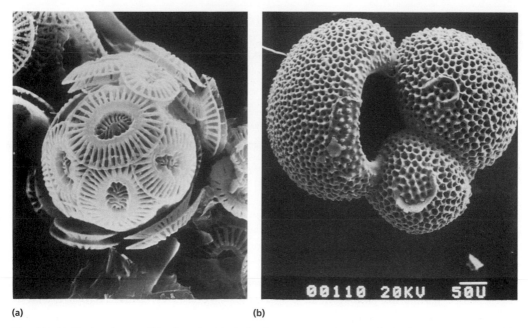

(a) **(b)**

Fig. 4.8 (a) Planktonic coccolithophore *Emiliania huxleyi,* a very common species in the modern oceans. This specimen has a diameter of 8 μm. The skeleton is clearly composed of circular shields packed around a single algal cell. After death the coccosphere breaks down, releasing the shields to form the microscopic particles of deep-sea oozes and chalks. Photograph courtesy of D. Harbour. (b) Modern planktonic foraminifer *Globigerinoides sacculifer*, common in tropical oceans. Scale bar = 50 μm. Photograph courtesy of B. Funnell.

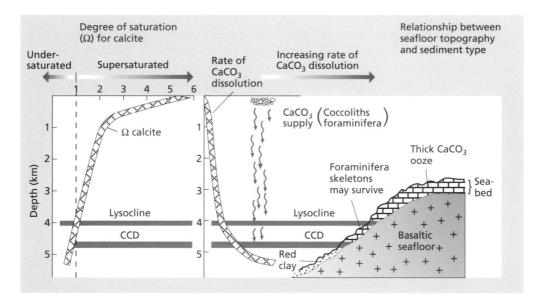

Fig. 4.9 Schematic diagram showing depth relationship between degree of saturation for calcite in seawater and rate of $CaCO_3$ dissolution. At 4 km depth, as seawater approaches undersaturation with respect to calcite, rate of dissolution of sinking calcite skeletons increases. The lysocline marks this increased rate of dissolution. Below the lysocline only large grains (foraminifera) survive dissolution if buried in the seabed sediment. Below the calcite compensation depth (CCD) — see text — all $CaCO_3$ dissolves, leaving red clays.

below 6 as such a shift requires a 1000-fold increase in atmospheric pCO_2 over its present value of $10^{-3.6}$ atm. An atmospheric pCO_2 of this magnitude may have occurred early in Earth history, but the existence of 3.8-billion-year-old carbonate sediments shows that seawater was still close to equilbrium with calcite, requiring a 10-fold increase in oceanic Ca^{2+} concentrations over modern values. Seawater pH has probably not exceeded 9, since sodium carbonate would then be much more common than calcite in ancient marine sediments. There is no evidence that sodium carbonate has ever been a common marine precipitate.

4.4.5 Opaline silica

Opaline silica (opal) is a form of biologically produced silicon dioxide ($SiO_2.nH_2O$) secreted as skeletal material by pelagic phytoplankton (diatoms) and one group of pelagic zooplankton (radiolarians) (Fig. 4.10). Opaline silica ($SiO_2.nH_2O$)-rich sediments cover about one-third of the seabed, mainly in areas where sedimentation rates are high, associated with nutrient-rich upwelling waters and polar seas, particularly around Antarctica (see Fig. 4.7). Seawater is undersaturated with respect to silica and it is estimated that 95% of opaline silica dissolves as it sinks through the water column or at the sediment/water interface. Thus, the preservation of opaline silica only occurs where it is buried in rapidly accumulating sediment, beneath the sediment/water interface. Subsequent dissolution of opal in the sediment saturates sediment pore waters with silica. The pore water cannot readily exchange with open seawater and saturation prevents further opal dissolution. High sedimentation rates in the oceans can be caused by high mineral supply rates from the continents, but are usually caused by high production rates of biological particles (see Section 4.5.2). In high productivity areas, diatoms are the common phytoplankton species, and this enhances the importance of these regions as silica sinks. The biological removal of silicon (Si) from seawater is calculated from the opal content of sediments and rates of sedimentation (see Table 4.2).

4.4.6 Sulphides

The oxidation of organic matter proceeds by a number of microbially mediated reactions once free oxygen has been used up (see Section 3.7.4). Although small amounts of nitrate (NO_3^-), manganese (Mn) and iron (Fe) are available as electron acceptors in marine sediments, their importance is small in comparison with SO_4^{2-}, which is abundant in seawater (see Table 4.1). At seawater pH around 8, sulphate-reducing bacteria metabolise organic matter according to the following simplified equation.

$$2CH_2O_{(s)} + SO_{4\ (aq)}^{2-} \rightarrow 2HCO_{3(aq)}^- + HS_{(aq)}^- + H_{(aq)}^+ \qquad \text{eq. 4.9}$$

This process is widespread in marine sediments but is most important in continental margin sediments, where organic matter accumulation is largest. Sulphate reduction in sediments occurs at depths (varying from a few millimetres to metres

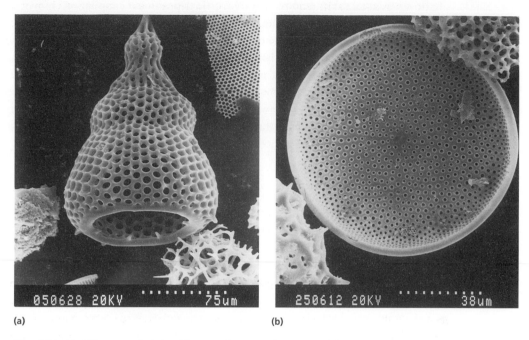

(a)   (b)

Fig. 4.10 (a) Siliceous radiolarian *Theocorythium vetulum,* early Pleistocene, equatorial Pacific. Scale bar = 75 µm. (b) Siliceous diatom *Coscinodiscus radiatus*, early Pleistocene, equatorial Pacific. Scale bar = 38 µm. Photographs courtesy of B. Funnell.

below the sediment/water interface) where seawater SO_4^{2-} can readily diffuse, or be pumped by sediment-dwelling organisms. The reaction yields highly reactive hydrogen sulphide (HS^-), most of which diffuses upward and is reoxidised to SO_4^{2-} by oxygenated seawater in the surface sediment. However, about 10% of the HS^- rapidly precipitates soluble Fe(II) in the reducing sediments to yield iron monosulphide (FeS).

$$Fe^{2+}_{(aq)} + HS^-_{(aq)} \rightarrow FeS_{(s)} + H^+_{(aq)}$$ eq. 4.10

Iron monosulphides convert to pyrite (FeS_2), given time; the process of conversion is poorly understood, but seems to involve the addition of sulphur (S) from intermediate sulphur species (e.g. polysulphides, polythionates or thiosulphate ($S_2O_3^{2-}$), which are the products and reactants in microenvironmental sulphur redox cycling. The reaction involving $S_2O_3^{2-}$ can be summarised as:

$$FeS_{(s)} + S_2O_{3(aq)}^{2-} \rightarrow FeS_{2(s)} + SO_{3(aq)}^{2-}$$ eq. 4.11

The sulphite (SO_3^{2-}) is subsequently oxidised to SO_4^{2-}. Sedimentary pyrite, formed as a by-product of sulphate reduction in marine sediments, is a major sink for seawater SO_4^{2-}. The presence of pyrite in ancient marine sediments shows that SO_4^{2-} reduction has occurred for hundreds of millions of years. On a geological timescale, removal of SO_4^{2-} from seawater by sedimentary pyrite formation is thought to be about equal to that removed by evaporite deposition. Compilations of pyrite abundance and accumulation rates are used to calculate modern SO_4^{2-} removal by this mechanism and to derive the estimate in Table 4.2.

Sulphate reduction (eq. 4.9) also produces HCO_3^- and this anion slowly diffuses out of the sediment into seawater, accounting for about 7% of the HCO_3^- flux to the oceans. The slow diffusion also means that HCO_3^- may build up to sufficiently high concentrations in sediment pore waters for the ion activity product of Ca^{2+} (from seawater) and HCO_3^- to exceed the solubility product for $CaCO_3$ (eq. 4.5). This allows $CaCO_3$ to precipitate as nodules (concretions) in the sediment.

4.4.7 Hydrothermal processes

The hydrothermal cycling of seawater through mid-ocean ridges (Box 4.7) changes the chemistry of some major and trace elements in the circulating seawater. Estimates of element fluxes in this process are uncertain, mainly because representative sampling in these remote environments is difficult and expensive. The flux estimates are based on a few studies at individual sites on the East Pacific and Atlantic ridges (Fig. 4.11). Global fluxes have been calculated from these sites, using various geophysical and geochemical approaches. A major problem, still to be resolved, is a rigorous quantification of the amount of hydrothermal activity occurring at high temperatures near the ridge axis, versus lower temperature circulation on the ridge flanks. This is important, since temperature affects the degree, rate and even direction of some chemical reactions. Despite these problems, for some elements the direction of the fluxes agree from site to site, which is encouraging. However, the magnitudes of the fluxes in Table 4.2 are uncertain.

Hydrothermal reactions as major ion sinks. Of the major elements, the case for magnesium removal from seawater during hydrothermal cycling at mid-ocean ridges is most convincing. Experimental work and data from the Galapagos hot springs (Fig. 4.11) suggested that hydrothermal fluids which exit from the crust have essentially zero magnesium concentration. This implies that magnesium is removed from seawater by reaction with basalt at high temperature. The precise chemistry of this reaction is not known, but it can be represented as:

$$11Fe_2SiO_{4(s)} + 2H_2O_{(l)} + 2Mg^{2+}_{(aq)} + 2SO^{2-}_{4(aq)} \rightarrow Mg_2Si_3O_6(OH)_{4(s)} + 7Fe_3O_{4(s)}$$

(fayalite) (seawater) (sepiolite) (magnetite)

$$+ FeS_{2(s)} + 8SiO_{2(aq)}$$

(pyrite) (silica) eq. 4.12

The basalt, represented here by iron-rich olivine (fayalite), is leached of its iron and hydrated by seawater, whilst Mg^{2+} from seawater is used to form the magnesium clay mineral (sepiolite), which represents altered basalt. The reaction also predicts the formation of iron oxide (Fe_3O_4) iron sulphide (FeS_2) and silica, all of which are found at hydrothermal vents and have been reproduced in the laboratory.

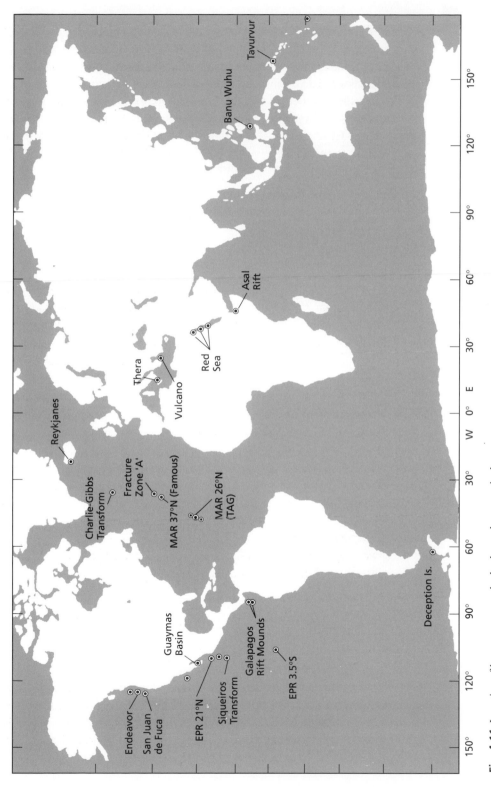

Fig. 4.11 Location of known seawater hydrothermal systems in the oceans.

Box 4.7

Hydrothermal circulation at mid-ocean ridges

Basaltic ocean crust is emplaced at mid-ocean ridges by crystallisation from magma emanating from magma chambers at shallow depth (about 2 km) below the ridge. The magma chamber and newly emplaced basalts can be viewed as a heat source, localised below the ridge (Fig. 1). Successive emplacement of new ocean crust gradually pushes the older crust laterally away from the ridge axis at rates of a few millimetres per year. This ageing crust cools and subsides as it travels away from the ridge axis. The resulting thermal structure, i.e. a localised heat source underlying the ridge with cooler flanking areas, encourages seawater to convect through fractures and fissures in the crust.

The deep waters of the oceans are cool (around 4°C) and dense relative to overlying seawater. This dense water percolates into fissures in the basaltic crust and approaches the heat source of the underlying magma chamber. This massive heat source warms the water, causing it to expand and become less dense, forcing it upward again through the crust in a huge convection cell (Fig. 2). We can view this convection cell as having two parts, a low-temperature 'limb' of subsiding seawater and a high-temperature rising 'limb' of chemically modified seawater. The overall process is often called 'hydrothermal' (hot water) convection.

It is not possible to measure directly the maximum temperature to which water becomes heated in the basaltic crust. However, hot springs of hydrothermal water issue from the seabed at the apex of the convection cell. Due to chemical reactions between the convecting water and the basaltic crust (shaded region on Fig. 2), these waters — which are acidified (typical pH 5–7) and rich in dissolved transition metals (iron (Fe), manganese (Mn), lead (Pb), zinc (Zn) and copper (Cu)) and hydrogen sulphides — rapidly precipitate a cloud of iron, zinc, lead and

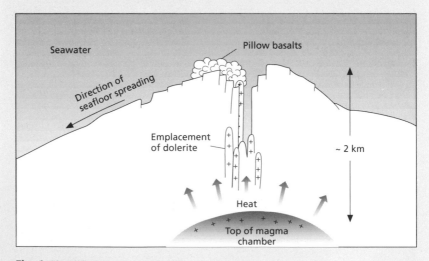

Fig. 1 Simplified structure of a mid-ocean ridge.

Box 4.7
Cont.

copper sulphides and iron oxides on injection into cold, oxic oceanic bottom waters. This sulphidic plume of particles identifies clearly the location of the hot springs and gives rise to their colloquial name — 'black smokers' (Fig. 3).

Temperature measurements taken from black smokers range from 200 to 400°C (average around 350°C). The high temperatures encountered in hydrothermal circulation cells at mid-ocean ridges increase substantially the rate and extent of chemical reaction between seawater and basaltic ocean crust. These reactions affect the global major ion budgets of magnesium, silicon, potassium and calcium and probably affect sodium and sulphate cycling.

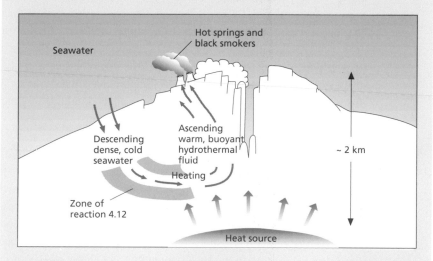

Fig. 2 Hydrothermal convection at a mid-ocean ridge.

Fig. 3 Black smoker, East Pacific Rise. From Des Océans aux continents. *BSGF*, 1984, (**7**), XXVI, no 3. Société Géologique de France, Paris. Reproduced with thanks.

Although it is difficult to quantify the amount of Mg^{2+} removed from seawater by this process, it is probably the most important Mg^{2+} sink in the modern ocean. There is also uncertainty about the fate of Mg^{2+} in altered basalt (sepiolite in eq. 4.12) as it moves away from the ridge axis during seafloor spreading. There is evidence that Mg^{2+} is leached from altered basalt by cold seawater. If large amounts of Mg^{2+} are resupplied to seawater by low-temperature basalt–seawater interaction, then mid-ocean ridge processes may not cause *net* Mg^{2+} removal from seawater.

If the case for net Mg^{2+} removal from seawater by basalt–seawater interactions is equivocal, the case for Na^+ removal is positively tenuous. Data to identify processes and quantify fluxes are scant. The main reason to expect removal of Na^+ from seawater at mid-ocean ridges is that the global Na^+ budget does not balance. The existence of Na^+-enriched basalts (spilites), believed to have formed by reaction with seawater at high temperature, is tangible evidence that Na^+ removal from seawater may occur during mid-ocean ridge hydrothermal activity. It is thought that hydrothermal activity accounts for no more than 20% of the total Na^+ removal from seawater, which on a geological timescale is dominated by the formation of evaporites (see Section 4.4.2).

Hydrothermal reactions as major ion sources. The chemistry of hydrothermal fluids indicates that basalt–seawater interactions are a source of some elements which are stripped from ocean crust and injected into seawater. Data from the Galapagos hot springs show that both Ca^{2+} and dissolved silica are concentrated in the hydrothermal waters compared with seawater (Table 4.4). Calcium is probably leached from calcium feldspars (anorthite), while silica can be leached from any decomposing silicate in the basalt, including the glassy matrix of the rock. If the Galapagos values are representative of average mid-ocean ridge hydrothermal element fluxes, then, globally, basalt–seawater interaction provides an additional 35% to the river flux of Ca^{2+} and silica to the oceans.

Hydrothermal reactions involving sulphate and potassium. The effects of hydrothermal reactions on oceanic SO_4^{2-} and K^+ budgets are even more problematic. Sulphur, either as hydrogen sulphide (H_2S) or HS^-, has been identified in hydrothermal fluids and the precipitation of SO_4^{2-} (anhydrite ($CaSO_4$)) is known to occur both in the crust from downward-percolating seawater and at vent sites. The fate of $CaSO_4$ in the crust is unknown, whilst the vent $CaSO_4$ probably redissolves in the ocean bottom waters and has little effect on the overall SO_4^{2-} budget. Earlier estimates of large-scale removal of SO_4^{2-} at mid-ocean ridges, based on data from the Galapagos site, are probably incorrect and seriously imbalance an otherwise satisfactory budget. It is well known that H_2S precipitates as iron sulphides from hydrothermal fluids, but the total removal of SO_4^{2-} from seawater by this mechanism is likely to be small, since evaporite and sedimentary sulphide formation adequately removes the river flux of SO_4^{2-} on geological timescales.

(a)

(b)

Plate 2.1 (a) Flying into Los Angeles in the early morning before an intense photochemical smog has developed. (b) Later in the day further north on the west coast (USA). Lengthy exposure of primary pollutants to sunlight has induced the photochemical formation of a brown smog. Photographs courtesy of P. Brimblecombe.

Facing P. 142

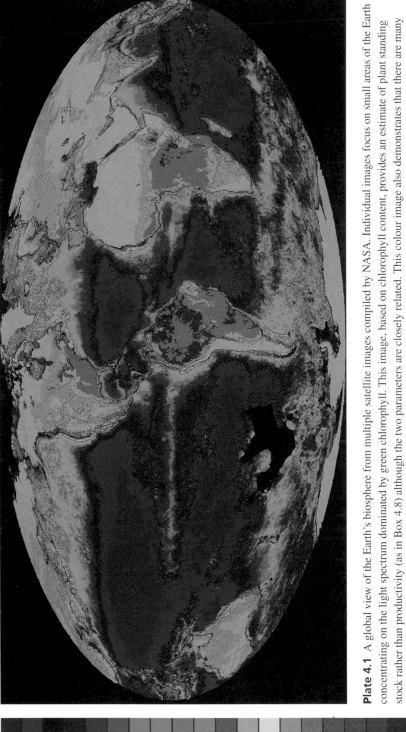

Plate 4.1 A global view of the Earth's biosphere from multiple satellite images compiled by NASA. Individual images focus on small areas of the Earth concentrating on the light spectrum dominated by green chlorophyll. This image, based on chlorophyll content, provides an estimate of plant standing stock rather than productivity (as in Box 4.8) although the two parameters are closely related. This colour image also demonstrates that there are many small-scale ocean features, such as eddies, which would complicate the simplified map in Box 4.8. The image also highlights the importance of tropical forests as areas of high plant biomass. Estimates of phytoplankton pigment concentrations (mg/m³) shown according to colour code.

Table 4.4 Concentration changes of some seawater major constituents upon reacting with basalt at high temperature. Data from Berner & Berner (1987)

Constituent	Seawater (mmol l^{-1})	Galapagos* (mmol l^{-1})	Δ† (mmol l^{-1})
Mg^{2+}	53	0	−53
Ca^{2+}	10	35	25
K^+	10	19	9
SO_4^{2-}	28	0	−28
H_4SiO_4	0.1	c. 20	c. 20
ΔCa^{2+} minus ΔSO_4^{2-}	—	—	53‡

* Galapagos data are extrapolated.
† Δ, difference between 350°C Galapagos hydrothermal water and seawater.
‡ The 'ΔCa^{2+} minus ΔSO_4^{2-}, term = the total Ca^{2+} leached from basalt — in other words, calcium leached from basalt but reprecipitated with sulphate, forming $CaSO_4$. This sulphate may redissolve as seafloor spreading moves material away from vent sites.

Laboratory studies predict that K^+ behaviour changes with temperature in hydrothermal fluids. Above 300°C, in the hottest part of hydrothermal systems, K^+ should be leached from basalt, representing an input to the seawater budget. However, in cooler parts (<300°C) of hydrothermal systems, K^+ adsorption on to altered basalt may be important, resulting in the formation of clay-like minerals such as celadonite (illitic) and phillipsite (a zeolite mineral). As there is no well-documented major removal process for K^+ from seawater, it is generally believed that low-temperature hydrothermal activity removes all of the high-temperature hydrothermal K^+ input to seawater and probably much of the river flux also.

4.4.8 Balancing the seawater major ion budget

The major ion budget for seawater (see Table 4.2) is quite well balanced (i.e. inputs equal outputs) for all elements except K^+. An uncertain amount of K^+ is probably consumed during low-temperature reactions between basalt and seawater, in addition to ion exchange reactions, which may remove up to 17% of the modern river flux of K^+ (see Table 4.3). The imbalance in the K^+ budget and small imbalances in other budgets may be nullified by a number of possible reactions between seawater and ocean sediments.

One possibility is 'reverse weathering reactions'. In reverse weathering, highly degraded clay minerals react with cations, HCO_3^- and silica in seawater to form complex clay mineral-like silicates. An example reaction addressing the K^+ problem would be:

$$\text{Degraded aluminosilicate}_{(s)} + K^+_{(aq)} + HCO^-_{3(aq)} + H_4SiO_{4(aq)} \rightarrow K \text{ aluminosilicate}_{(s)} + CO_{2(g)} + H_2O_{(1)} \qquad \text{eq. 4.13}$$

This reaction shows that reverse weathering is exactly opposite to continental weathering reactions, which consume CO_2 and liberate HCO_3^- (see Section 3.4.3). The concept of reverse weathering has proved difficult to corroborate by observational data. Limited data suggest that the process may occur in Atlantic Ocean sediments, but definitive data are needed.

Another process which might affect the K^+ budget in a small way is K^+ fixation during ion exchange reactions on clay minerals. Laboratory experiments have shown that degraded micas and illites (see Section 3.6.4), stripped of their K^+ during weathering, but which retain much of their layer charge, are able to fix, irreversibly, K^+ from seawater. The process involves the replacement of hydrated cations for dehydrated K^+ in the interlayer site, fixing the K^+ in its 'mica' site (see Section 3.6.4). Globally, this process might remove about 10% of the K^+ river flux to the oceans.

Some removal of ions from seawater occurs through permanent burial in sediment pore water. The total removal of major ions by this process is small, less than 2% of the river input for all elements except Na^+ and Cl^-. The burial flux may be significant for Na^+ and Cl^- (20–30% of the river flux), but the data are uncertain.

Seawater buried in marine sediment may react with components of the sediment, particularly fine-grained basaltic volcanic ash. Pore water concentrations of Ca^{2+}, Mg^{2+} and K^+ in deep-sea cores show removal of Mg^{2+} (and, to a lesser extent, K^+) from pore water, mirrored by increases in Ca^{2+} pore water concentration (Fig. 4.12). These results suggest that basaltic ash is converted to Mg^{2+} and K^+ clay minerals, accompanied by the release of Ca^{2+} to pore water. The quantitative importance of this mechanism on a global scale is probably small, but good data are sparse.

4.4.9 Anthropogenic effects on major ions in seawater

A final complication to major ion budgets in seawater is that the estimated river input to seawater is that considered to be naturally derived. Anthropogenic processes have altered some of these fluxes. For example, the riverine Cl^- flux may have increased by more than 40% as a result of human activity and the SO_4^{2-} flux may have doubled, mainly due to fossil fuel combustion and oxidation of pollution-derived H_2S.

Human activity also affects the volume of suspended load carried by rivers. Deforestation and increased agricultural activity worldwide make land surfaces more susceptible to erosion. The effects on the major ion chemistry of seawater are mainly related to increased input of detrital solids to continental shelves, which increases the amount of ion exchange and other solid–seawater interactions. However, this situation is still changing; the increasing use of dams on rivers is now reducing sediment inputs to the oceans. Less important in terms of global budgets, but important from an ecological viewpoint, increased suspended loads in tropical rivers issuing into coastal waters choke coral reefs with detritus and decrease biological productivity by reducing water clarity.

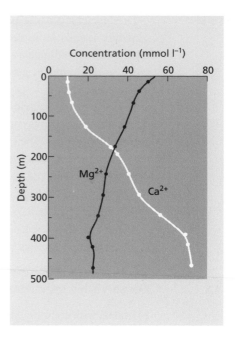

Fig. 4.12 Depth distribution of dissolved Ca^{2+} and Mg^{2+} concentrations in sediment pore waters. After *Geochimica Cosmochimica Acta*, **45**, Gieskes, J.M. & Lawrence, J.R. Alteration of volcanic matter in deep-sea sediments: evidence from the chemical composition of interstitial waters from deep-sea drilling cores,1694. Copyright 1981, with kind permission from Pergamon Press Ltd, Headington Hill Hall, Oxford 0X3 OBW, UK.

4.5   ## Minor chemical components in seawater

Seven major ions dominate the chemistry of seawater, but all of the other elements are also present, albeit often at extremely low concentrations. The major ions in seawater are little affected by biological processes or human activities because seawater is a vast reservoir and the major ions have long residence times. In contrast, complex cycling processes and involvement in biological systems typify the behaviour of dissolved trace elements (components present at µmol l^{-1} concentrations or less) in seawater. The concentration of some dissolved metals in seawater is very small — typically a few nmol l^{-1}. Sampling so as to avoid contamination and measuring such tiny concentrations, in the presence of major ions with millimolar concentrations, is difficult. These difficulties prevented routine analysis of trace metals in seawater until the 1970s, although reliable nutrient measurements were available earlier.

Particulate matter concentrations in the deep ocean are low (a few µg l^{-1}), whereas in surface waters particulate matter concentrations are relatively high (generally 10–100 µg l^{-1}). Similarly high values are encountered close to the deep ocean floor, caused by the resuspension of deep-sea sediments, and locally (tens of kilometres) around hydrothermal vents (Box 4.7). Apart from this region near the seafloor, particulate matter in the oceans is predominantly of organic origin, generated by primary production in surface seawater. The euphotic zone, where this production occurs, has variable depth, generally around 100 m in clear open-ocean waters. Since the oceans are on average almost 4000 m deep, the primary production which drives global biological cycling occurs in a shallow surface zone.

Dissolved metals in seawater have various sources, e.g. the dissolution of redox-sensitive metals from reducing ocean floor and mid-ocean ridge sediments. Hydrothermal sediments are often manganese-rich, although the restricted occurrence around vents means that the contribution to the global dissolved manganese budget is uncertain. In general, atmospheric and riverine inputs are most important. Modern atmospheric fluxes of some metals are larger than river inputs (Table 4.5), caused by various combustion processes — coal burning, metal smelting and automobile engines. The shift toward a larger atmospheric source for some metals may increase their concentrations in open-ocean waters, since riverine metal inputs are often removed in estuaries (see Section 4.2).

The chemistry of dissolved metals in seawater can be grouped into three classes, which describe the behaviour of the metal during chemical cycling. These classes — conservative, nutrient-like and scavenged — have been recognised by the shapes of concentration profiles when plotted against depth in the oceans.

4.5.1 Conservative behaviour

Elements with conservative behaviour are characterised by vertical profiles (similar to major ion profiles) that indicate essentially constant concentrations with depth. These elements behave like the major ions, having long residence times and being well mixed in seawater. These elements are not major components of seawater, simply because their crustal abundances are very low compared with the major ions. Elements showing this sort of behaviour form simple anions or cations (with low z/r ratios and hence little interaction with water), e.g. Cs^+ or bromine ion (Br^-), or form complex oxyanions, e.g. molybdenum (Mo) and tungsten (W) — which

Table 4.5 Comparison of total atmospheric and riverine inputs to the world's oceans (10^9 mol year^{-1}). Based on Duce *et al.* (1991)

Element	Riverine*	Atmospheric†
Nitrogen (ex N_2)	1500–3570	2140
Cadmium	0.0027	0.02–0.04
Copper	0.16	0.25–0.82
Nickel	0.19	0.37–0.48
Iron	19.7	57
Lead	0.01	0.43
Zinc	0.09	0.67–3.5

* Dissolved input only; particulate components are assumed to sediment out in estuaries and the coastal zone.
† Total input — dissolved plus particulate.
Estimates are based on data available in early 1990s and so include significant amounts of material mobilised by human activity.

exist in seawater as MoO_4^{2-} and WO_4^{2-} respectively (Fig. 4.13). Conservative elements have little interaction with biological cycles.

4.5.2 Nutrient-like behaviour

As in continental waters (see Section 3.7.5), NO_3^-, DIP and silicate are usually considered to be the limiting nutrients for biological production, although in some situations it has been suggested that trace elements, particularly iron, may be limiting. The oceans are so large and deep that they are effectively permanently stratified. The production of biological material removes nutrients from surface waters (Box 4.8). After death, this biological material sinks through the water column, decomposing at depth to re-release the nutrients. The nutrients are then slowly returned to surface waters by deep-ocean mixing processes and diffusion. The net result is that the vertical profiles of nutrients are characterised by low concentrations in surface waters (where biological utilisation rates exceed supply rates) and deep-water maxima, where decomposition rates exceed uptake rates because of the absence of light (Fig. 4.14). Nitrogen and phosphorus are cycled

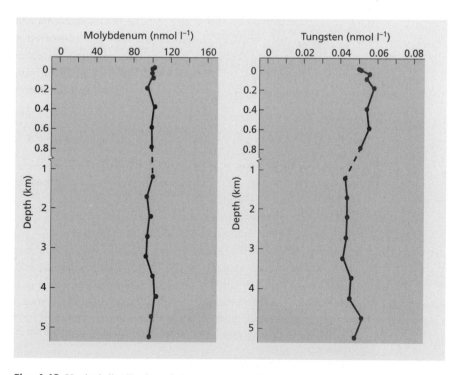

Fig. 4.13 Vertical distribution of dissolved molybdenum and tungsten in the North Pacific. After Sohrin *et al.* (1987).

Box 4.8

Oceanic primary productivity

The rate of growth of phytoplankton (primary productivity) in the oceans is mainly limited by the availability of light and the rate of supply of limiting nutrients (usually accepted to be nitrogen (N), phosphorus (P) and silicon (Si)). The need for light confines productivity to the upper layers of oceans. Also, in polar waters there will be no phytoplankton growth during the dark winter months.

In temperate oceans there is little winter productivity because cooling of surface waters destroys the thermal stratification and winds mix waters and phytoplankton to depths of hundreds of metres. This deep mixing, which occurs in all temperate and polar oceans, also means that vertical gradients in nutrient concentrations are temporarily eliminated, allowing a corresponding rise in surface-water nutrient concentrations. In spring, surface-water stratification is re-established as the waters warm and winds decrease. Once enough light is available, vigorous plankton growth begins. Nutrient concentrations are high, having built up over the winter, resulting in the 'spring bloom' of phytoplankton. The bloom is generally later in the spring — and larger — moving polewards.

The duration of the spring bloom is limited by nutrient availability and/or grazing by zooplankton. Phytoplankton growth and abundance then declines to lower levels, which are maintained throughout the summer. In some locations, limited mixing in autumn can stimulate another small bloom, before deep winter mixing returns the system to its winter condition.

In tropical waters, vertical stratification persists throughout the year and production is permanently limited by nutrient supply rates, which are controlled by internal recycling and slow upward diffusion from deep water. Under these conditions, productivity is low throughout the year.

These seasonal cycles of productivity are shown schematically in Fig. 1, along with a map of current estimates of primary production rates. In the north Pacific zooplankton population growth supresses the spring bloom.

Since production rates vary with time and place, the data on the figure are uncertain, but are consistent with satellite-derived maps of chlorophyll concentrations (see Plate 4.1, facing p. 143). These maps show that, on an annual basis, the short, high-production seasons of temperate and polar areas fix more carbon in organic tissue than organisms in tropical waters. There are a few exceptions to this, in so-called upwelling areas, e.g. the Peruvian, Californian, Namibian and North African coasts and along the line of the equator (Plate 4.1 and Fig. 1). In upwelling areas, ocean currents bring deep water to the surface, providing a large supply of nutrients in an area with abundant light. Very high rates of primary production ensue and the phytoplankton are the basis of a food chain that supports commercially important fisheries.

The spring bloom and upwelling areas are not simply times and regions of higher productivity; changes in the structure of the whole ecosystem result. For example, the phytoplankton community in areas of higher production is usually rich in diatoms, organisms which, upon death, efficiently export carbon and nutrients to deep waters. This contrasts with the tropics, where the phytoplankton

Box 4.8 Cont.

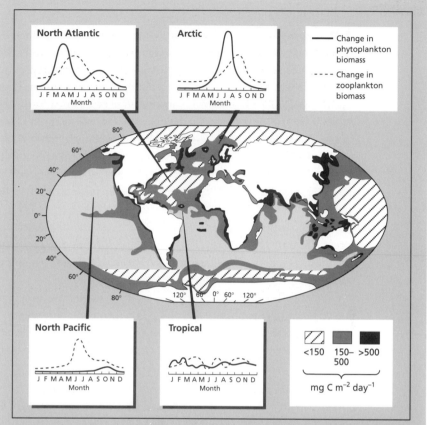

Fig. 1 Seasonal cycles of productivity and global average rates of primary production. Reprinted with permission from *Scientific Exploration of the South Pacific*, Courtesy of the National Academy of Sciences, Washington, D.C.

community has adapted to the low-nutrient waters by recycling nutrients very efficiently, with little export to deep waters.

with the organic tissue of organisms, while silicon and calcium (Ca) are cycled as skeletal material. The decomposition of organic tissue is mainly by bacterial respiration, a rapid and efficient process. In contrast, skeletal material is dissolved slowly (see Sections 4.4.4 and 4.4.5). The effect of these different decomposition rates is that the NO_3^- and phosphorus concentration profiles show rapid increase with depth, implying shallower regeneration of material in the water column than silicon.

Biological cycling not only removes nutrients from surface waters, it also transforms them. The stable form of iodine (I) in seawater is iodate (IO_3^-) but biological cycling results in the formation of iodide (I^-) in surface waters, because the production rate of the reduced species exceeds the oxidation rate. Biological uptake of IO_3^- in surface water results in nutrient-like, rather than conservative, behaviour. The biological demand for NO_3^- also involves transformation. Phytoplankton take up NO_3^- and reduce it to the -3 oxidation state (see Box 3.5) for utilisation in proteins. When phytoplankton die they decompose, releasing the nitrogen as ammonium (NH_4^+) hence N still -3. Similarly, when phytoplankton are eaten by zooplankton, the latter organisms excrete nitrogen primarily as NH_4^+. This NH_4^+ is then available for reuse by phytoplankton: NH_4^+ is the preferred form of available nitrogen since there is no energy requirement in its uptake and utilisation. Alternatively, the ammonium is oxidised via nitrite (NO_2^-) to NO_3^-. These rapid recycling processes maintain euphotic zone NH_4^+ at very low concentrations. In the deep ocean, the only NH_4^+ source is from the breakdown of organic matter sinking from the surface waters. The amount of NH_4^+ released from this source is small and rapidly oxidised, maintaining low NH_4^+ concentrations.

Elements showing nutrient-like distribution often have very long oceanic residence times. The residence times of NO_3^- silicon and phosphorus have been estimated to be 57 000, 20 000 and 69 000 years respectively (Table 4.6). The vast reservoirs of nutrients in the deep ocean mean that increases in the concentrations

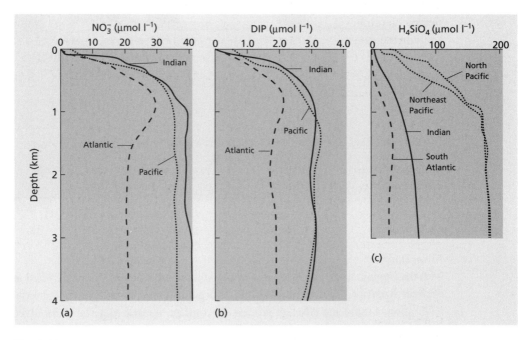

Fig. 4.14 Vertical distribution of dissolved nitrate (a), phosphorus (b) and silicon (c) in the Atlantic, Pacific and Indian Oceans. After Svedrup *et al.* (1941).

of NO_3^- in riverwaters due to human activity (see Section 3.7.5) have little effect on oceanic NO_3^- concentration (this assumes that NO_3^- is effectively mixed throughout the ocean volume). We will consider the validity of this assumption later, but first we can illustrate the case for complete mixing.

The average NO_3^- concentration in the oceans is 30 µmol l^{-1}. The total oceanic NO_3^- inventory is calculated by multiplying this concentration by the volume of the oceans, i.e.

$$30 \times 10^{-6} \ (\text{mol } NO_3^- \ l^{-1}) \times 1.37 \times 10^{21} \ (l) = 41 \times 10^{15} \ \text{mol } NO_3^- \qquad \text{eq. 4.14}$$

It is difficult to estimate the natural NO_3^- concentration in rivers since most have been affected by human activities to some extent. Nitrate concentrations in the Amazon are about 20 µmol l^{-1} and are probably close to natural levels. Multiplying this concentration by the global river flux gives an estimate of the natural nitrate input to the oceans.

$$20 \times 10^{-6} (\text{mol } NO_3^- l^{-1}) \times 3.6 \times 10^{16} \ (l \ year^{-1}) = 0.72 \times 10^{12} \ \text{mol } NO_3^- \ year^{-1}$$
$$\text{eq. 4.15}$$

The NO_3^- concentrations in some rivers have been increased to concentrations as high as 500 µmol l^{-1}, although the greatest changes have generally been in small rivers which make a small contribution to the total river flux. If, as an extreme case, we suppose that 10% of the world river flow increased to this high concentration, we can calculate how long such a situation would take to double the oceanic NO_3^- concentration, i.e. introduce 41×10^{15} mol. The calculation is simplified by ignoring NO_3^- removal to sediment. In reality, increased NO_3^- inputs would result in increased biological activity and increased NO_3^- removal from the oceans, since the main sink for nitrogen and phosphorus is burial in organic matter. If 10% of the world's rivers have NO_3^- concentrations increased to 500 µmol l^{-1}, total riverine NO_3^- inputs become:

Table 4.6 Concentrations of nutrients and metals in deep (> 3000 m) water in the North Atlantic and North Pacific, together with estimated oceanic residence times

Component	North Atlantic	North Pacific	Estimated oceanic residence time (years)
Nitrate (µmol l^{-1})	20	40	57 000
Silicon (µmol l^{-1})	25	170	20 000
Phosphorus (µmol l^{-1})	1.3	2.8	69 000
Zinc (nmol l^{-1})	1.7	8.0	4 500
Cadmium (nmol l^{-1})	0.3	0.9	32 000
Aluminium (nmol l^{-1})	20	0.4	50
Manganese (nmol l^{-1})	0.6	0.2	30

$[20 \times 10^{-6} \, (\text{mol NO}_3^- \text{l}^{-1}) \times 3.6 \times 10^{16} \, (1 \, \text{year}^{-1}) \times 0.9]$
$+ \, [500 \times 10^{-6} \, (\text{mol NO}_3^- \text{l}^1) \times 3.6 \times 10^{16} \, (1 \, \text{year}^{-1}) \times 0.1]$
$= 2.45 \times 10^{12} \, \text{mol NO}_3^- \, \text{year}^{-1}$

eq. 4.16

The time needed to double the oceanic nitrate inventory, assuming no removal at all (hence a minimum estimate), is then:

$$\frac{41 \times 10^{15} \, \text{mol NO}_3^-}{2.45 \times 10^{12} \, \text{mol NO}_3^- \, \text{year}^{-1}} = 16\,700 \, \text{years}$$

eq. 4.17

Thus, even drastic perturbations of the freshwater NO_3^- input to the oceans cannot alter the seawater concentration rapidly because of the huge oceanic reservoir of this element. We should note, however, that in smaller regional seas, such as the Baltic, the situation can be very different (Box 4.9).

In addition to the actual nutrient elements, many other elements show nutrient-like behaviour in the oceans, i.e. low concentrations in surface waters and high concentrations at depth (Fig. 4.15). This distribution implies that biological removal rates from surface waters are rapid, although it does not prove that these elements are limiting, or even essential, to biological processes. In the case of some metals (e.g. zinc (Zn)), a clear biological function has been established. However, for other metals (e.g. cadmium (Cd)), there is little evidence for a biological role; cadmium is usually thought of as a poison, although not at the extremely low concentrations (< 0.1 nmol l^{-1}) found in seawater. Elements like cadmium probably show nutrient-like behaviour (Fig. 4.15) because they are inadvertently taken up during biological processes. The Cd^{2+} ion has chemical similarities to Zn^{2+}, thus the nutrient-like cycling of cadmium may reflect inadvertent biological uptake in association with zinc.

Finally, we should be aware that even metals with a clear biological role (e.g. Zn^{2+}) can be toxic at sufficiently high concentrations. This reminds us that all elements are potentially toxic, making terms like *nutrient* impossible to apply in an absolute sense.

4.5.3 Scavenged behaviour

Elements that are highly particle-reactive, characterised by large z/r ratios (see Section 3.7.1), often have vertical profiles with surface maxima and decreasing concentrations with depth (Fig. 4.16). These profiles ̇se because the inputs of these elements are all in the surface waters, producing concentration maxima there. Poorly understood processes lower these concentrations by removal to particulate phases. The removal processes probably involve adsorption on to particle surfaces, known by the general term *scavenging*. Consequently, oceanic concentrations of scavenged elements are well below those predicted from simple mineral solubility considerations. The iron- and manganese-rich sediments precipitated from black smokers cause intense scavenging of some metals from seawater, making mid-ocean ridge axis environments sinks for these metals.

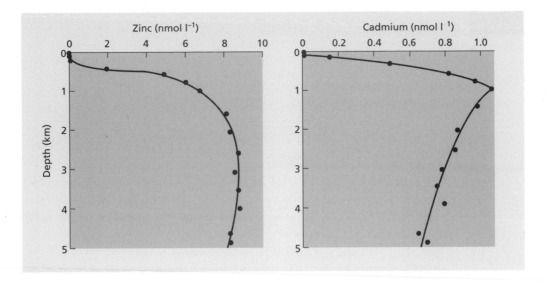

Fig. 4.15 Vertical distribution of dissolved zinc and cadmium in the North Pacific. After Bruland (1980).

Scavenged species are all metals and their residence times in seawater are estimated to be a few hundred years, short in comparison with nutrient and conservative elements (Table 4.6). These rapid removal rates mean that river inputs are removed largely by estuarine processes, where suspended solid concentrations are high (see Section 4.2.1). Consequently, the atmosphere provides the main input of particle-reactive metals to the surface waters of the central ocean. This atmospheric flux has a natural component, the fallout of wind-blown dust particles, which subsequently dissolve in seawater to a small extent (typically a few per cent).

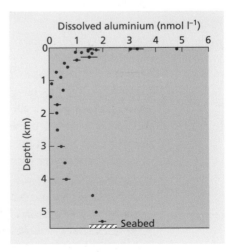

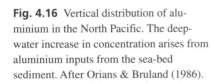

Fig. 4.16 Vertical distribution of aluminium in the North Pacific. The deepwater increase in concentration arises from aluminium inputs from the sea-bed sediment. After Orians & Bruland (1986).

Box 4.9

Human effects on regional seas: the Baltic

The Baltic Sea (Fig. 1) is a large regional sea, receiving drainage from much of northern and central Europe. The hydrography of the Baltic is complex, consisting of a number of deep basins separated by shallow sills. As a result, the waters of the deep basins can be isolated from exchange with one another — and from the atmosphere — on timescales of years.

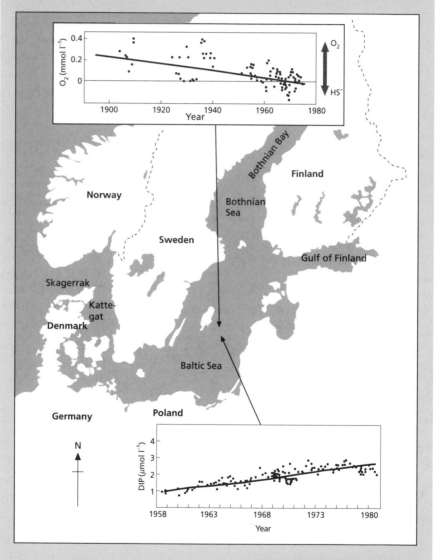

Fig. 1 Oxygen and phosphorus concentrations in the Baltic Sea. Dark line through data is a regression line and thin line is O_2 concentration. Reprinted from *Marine Pollution Bulletin*, **12**, Fonselius, S., Oxygen and hydrogen sulphide conditions in the Baltic Sea, 187–194 and Nehring, D., Phosphorus in the Baltic Sea, 194–198, Copyright 1981, with kind permission from Elsevier Science Ltd, The Boulevard, Langford Lane, Kidlington, OX5 1GB, UK.

Box 4.9
Cont.

There is a long record (almost 100 years) of dissolved phosphorus (P) and oxygen (O) concentrations for the waters of the Baltic. The records are 'noisy' due to complex water exchange and deep mixing, but the increasing concentration of dissolved phosphorus over the last 30 years is clear from Fig. 1. This increase in nutrient concentration has fuelled primary production and has increased the flux of organic matter to the deep waters. Measurements of dissolved oxygen in deep waters of the Baltic show a steady decline over the last 100 years (Fig. 1), consistent with an increase in rates of oxygen consumption due to increasing organic matter inputs — overall, a clear example of eutrophication.

The isolated deep waters of the Baltic have probably always had low oxygen concentrations. However, the declining trend over recent years means that, in some areas, oxygen concentrations have fallen to zero (anoxic). Under anoxic conditions, respiration of organic matter by microbial sulphate (SO_4^{2-}) reduction has produced hydrogen sulphides (HS^-) (plotted as negative oxygen in Fig. 1 above).

The Baltic contrasts with the nearby North Sea, where oxygen levels rarely fall, despite large inputs of nutrients. This is because the North Sea is shallow and its waters exchange freely with those of the North Atlantic.

Aluminium (Al) (Fig. 4.16) and iron are examples. The second source of particles is human activity: lead (Pb) is an example, entrained into the atmosphere principally from automobile exhaust emissions. Lead use, particularly as a petrol additive, increased rapidly during the 1950s until concern over the possible health effects resulted in a dramatic decline in its use from the late 1970s onward.

We do not have a direct history of dissolved lead concentrations in seawater but we do have an indirect record from corals. Coral skeletons are made of annual layers of $CaCO_3$, producing growth rings similar to those in trees. These rings can be counted and sampled for lead analysis. The lead ion, Pb^{2+}, is almost the same size and charge as Ca^{2+} and substitutes for it in the $CaCO_3$ coral skeleton, faithfully documenting the history of lead concentrations in surface seawater (Fig. 4.17).

4.5.4 Ocean circulation and its effects on trace element distribution

The preceding discussion of trace elements in seawater has assumed that the oceans have a uniform, warm, nutrient-depleted surface mixed layer and a static deep zone. In fact, at high latitudes the ocean surface waters are cold enough to destroy the density stratification and to mix the oceans to depths of up to 1000 m. This dense surface water sinks and flows slowly into the centre of the oceans as a layer of cold, oxygen-rich water, which displaces the bottom waters. The displaced bottom water is forced to move upwards slowly, setting up an oceanic circulation (Figs 4.18 and 4.19).

The deep mixing at high latitudes only occurs in two locations: in the North Atlantic and around Antarctica. Deep mixing does not occur in the North Pacific, mainly because a physical sill, related to the Aleutian Arc, prevents water exchange between the Arctic and the Pacific (Fig. 4.19). This asymmetry in deep mixing drives a global ocean circulation, in which surface water sinks in the North Atlantic, returns to the surface in the Antarctic and then sinks again and enters the Pacific and Indian Oceans (Fig. 4.18). The deep flow tends to concentrate at the

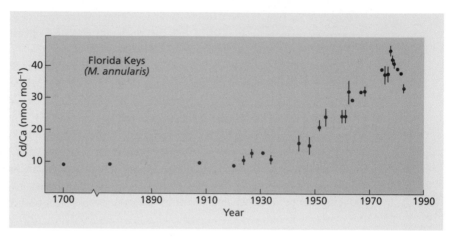

Fig. 4.17 Lead concentrations in dated year bands from a coral collected from the Florida Keys. After Shen & Boyle (1987).

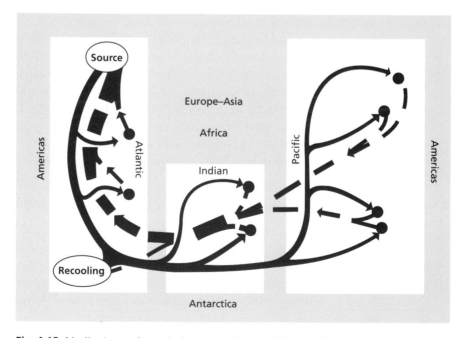

Fig. 4.18 Idealised map of oceanic deep-water flow (solid lines) and surface-water flow (dashed lines). Open circles represent areas of water sinking and dark circles of upwelling. After Broecker & Peng (1982).

western edge of ocean basins, but allows a slow diffusion of water throughout the ocean interiors. This slow, deep-water flow is compensated by a poleward return flow in surface waters (Fig. 4.20). A 'parcel' of seawater takes hundreds of years to complete this global ocean journey, during the course of which the deep water continually acquires the decay products of sinking organic matter from surface seawater. Waters in the North Pacific take longer to acquire these decay products since they are the 'oldest', in the sense of time elapsed since they were last at the surface and had their nutrients removed by biological processes. Similarly, the waters of the North Pacific have the lowest dissolved oxygen concentrations and high dissolved CO_2 concentrations, since oxygen has been used to oxidise greater amounts of organic matter. Overall, the supply of dissolved oxygen to seawater is adequate to oxidise the sinking organic matter and, apart from a few unusual areas in the oceans, oxygen concentrations in the bottom waters are adequate to support animal life. The higher dissolved CO_2 concentration in the Pacific results in a shallower CCD in the Pacific, relative to the Atlantic Ocean (see Section 4.4.4.).

The slower regeneration of silicon compared with nitrogen and phosphorus (see Section 4.5.2) means that relatively more silicon is regenerated in deep waters, producing steeper interocean concentration gradients (Table 4.6, Fig. 4.14). Similarly, other elements which show nutrient-like behaviour, such as zinc and cadmium, have higher concentrations in the North Pacific compared with other oceans. In contrast, scavenged elements have concentrations which are lower in the deep waters of the Pacific compared with the North Atlantic, because of the longer time

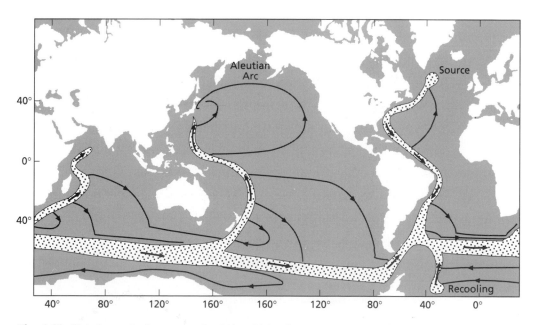

Fig. 4.19 Global oceanic deep-water circulation. Major flow routes are marked by stippled ornament. Deep mixing in the North Pacific is prevented by the topography of the seabed around the Aleutian Arc. Reprinted from Stommel, H. (1958) *Deep Sea Research*, **5**, 80–82, Pergamon, Oxford.

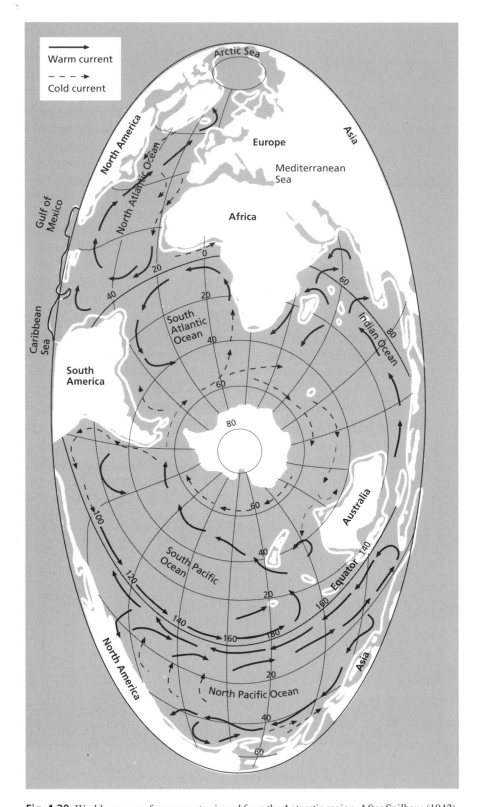

Fig. 4.20 World ocean-surface currents viewed from the Antarctic region. After Spilhaus (1942).

available for their removal by adsorption on to sinking particulate matter (Table 4.6).

This pattern of global oceanic circulation has probably existed since the end of the last glaciation, 11 000 years ago. Before this, the circulation pattern is thought to have been different, due to changes in glacial climatic regime and changes in polar ice volume. It is unclear whether changes in ocean circulation provoked climatic change at this time or vice versa. Despite the uncertainty, it is clear that ocean circulation and global climate are intimately linked.

4.6   Further reading

Berner, K.B. & Berner, R.A. (1987) *The Global Water Cycle*. Prentice Hall, Englewood Cliffs, NJ, 397 pp.

Broeker, W. & Peng, T.-H. (1982) *Tracers in the Sea*. Lamont Doherty Geological Observatory, Palisades, New York, 690 pp.

Chester, R. (1990) *Marine Geochemistry*. Unwin Hyman, London, 698 pp.

Drever, J.I., Li, Y.-H. & Maynard, J.B. (1988) Geochemical cycles: the continental crust and oceans. In: *Chemical Cycles in the Evolution of the Earth*, ed. by Gregor, C., Garrels, R.M., Mackenzie, F.T. & Maynard, J.B., pp. 17–53. Wiley, New York.

Libes, S. (1992) *Marine Biogeochemistry*. Wiley, New York, 735 pp.

Turner, B., Clark, W., Kates, R., Richards, J., Mathews, J. & Meyer, W. (1990) *The Earth as Transformed by Human Action*. Cambridge University Press, Cambridge, 713 pp.

5 Global change

5.1 Why study global-scale environmental chemistry?

In previous chapters of this book the chemistry of the atmosphere, oceans and land has been dealt with individually. Using a steady-state model (see Section 2.3), we can envisage each of these environments as a reservoir. In each chapter the cycling of chemicals has been discussed, together with their transformations within the reservoir; where relevant, some attention has been paid to inputs and outputs into or out of that reservoir from or to adjacent ones. In contrast, this chapter focuses, not on individual reservoirs, but on the ensemble of them that make up an integrated system, of air, water and solids, constituting the near-surface environments of our planet.

As scientists have learnt more about the way chemical constituents of the Earth's surface operate, it has become clear that it is insufficient to consider only individual environmental reservoirs. These reservoirs do not exist in isolation — there are large and continuous flows of chemicals between them. Furthermore, the outflow of material from one reservoir may have little effect on it, but can have a very large impact on the receiving reservoir. For example, the natural flow of reduced sulphur gas from the oceans to the atmosphere has essentially no impact on the chemistry of seawater, and yet has a major role in the acid–base chemistry of the atmosphere, as well as affecting the amount of cloud cover.

Since integrated systems need to be understood in a holistic way, studies of the global environmental system and natural and human-induced changes to it have become very important. By definition, such studies are on a large scale, generally beyond the resources of most nations, let alone individual scientists. Thus, in the last decade several large international programmes have been put in place, the most relevant to environmental chemistry being the International Geosphere–Biosphere Programme (IGBP) of the International Council of Scientific Unions. This has as its aim:

> To describe and understand the interactive physical, chemical and biological processes which regulate the total Earth system, the unique environment that it provides for life, the changes which are occurring in this system, and the manner in which they are influenced by human activities.

This large research agenda is concerned not only with understanding how Earth systems currently operate, but also with predicting how they may change in the future as a result of human activities and other factors.

5.2 What sorts of substances are involved?

In discussing the chemical aspects of global change it is important to distinguish